Springer Theses

Recognizing Outstanding Ph.D. Research

Aims and Scope

The series "Springer Theses" brings together a selection of the very best Ph.D. theses from around the world and across the physical sciences. Nominated and endorsed by two recognized specialists, each published volume has been selected for its scientific excellence and the high impact of its contents for the pertinent field of research. For greater accessibility to non-specialists, the published versions include an extended introduction, as well as a foreword by the student's supervisor explaining the special relevance of the work for the field. As a whole, the series will provide a valuable resource both for newcomers to the research fields described, and for other scientists seeking detailed background information on special questions. Finally, it provides an accredited documentation of the valuable contributions made by today's younger generation of scientists.

Theses are accepted into the series by invited nomination only and must fulfill all of the following criteria

- They must be written in good English.
- The topic should fall within the confines of Chemistry, Physics, Earth Sciences, Engineering and related interdisciplinary fields such as Materials, Nanoscience, Chemical Engineering, Complex Systems and Biophysics.
- The work reported in the thesis must represent a significant scientific advance.
- If the thesis includes previously published material, permission to reproduce this must be gained from the respective copyright holder.
- They must have been examined and passed during the 12 months prior to nomination.
- Each thesis should include a foreword by the supervisor outlining the significance of its content.
- The theses should have a clearly defined structure including an introduction accessible to scientists not expert in that particular field.

More information about this series at http://www.springer.com/series/8790

Snir Gazit

Dynamics Near Quantum Criticality in Two Space Dimensions

Doctoral Thesis accepted by
the Technion – Israel Institute of Technology,
Haifa, Israel

Author
Dr. Snir Gazit
Physics Department
University of California at Berkeley
Berkeley, CA
USA

Supervisor
Prof. Assa Auerbach
Physics Department
Technion – Israel Institute of Technology
Haifa
Israel

ISSN 2190-5053 ISSN 2190-5061 (electronic)
Springer Theses
ISBN 978-3-319-37014-9 ISBN 978-3-319-19354-0 (eBook)
DOI 10.1007/978-3-319-19354-0

Springer Cham Heidelberg New York Dordrecht London

Softcover reprint of the hardcover 1st edition 2015

Printed on acid-free paper

Springer International Publishing AG Switzerland is part of Springer Science+Business Media (www.springer.com)

Parts of this thesis have been published in the following journal articles:

1. Fate of the Higgs Mode Near Quantum Criticality, S. Gazit, D. Podolsky and A. Auerbach, Phys. Rev. Lett. 110, 140401 (2013).
2. Dynamics and Conductivity Near Quantum Criticality, S. Gazit, D. Podolsky, A. Auerbach and D. P. Arovas, Phys. Rev. B 88, 235108 (Editors' Suggestions, 2013).
3. Critical Capacitance and Charge-Vortex Duality Near the Superfluid-to-Insulator Transition, S. Gazit, D. Podolsky and A. Auerbach, Phys. Rev. Lett. 113, 240601 (2014).

Supervisor's Foreword

Recent beautiful experiments from the groups of Immanuel Bloch and Tilman Esslinger have discovered the superfluid to Mott insulator transition in optically trapped arrays of cold atoms. This transition realizes the quantum critical point (QCP) of the two-dimensional relativistic O(2) model. A disordered version of this QCP was realized much earlier in studies of the superconductor to insulator transition (SIT) in granular superconducting films. The SIT phenomenon has spurred enormous theoretical and experimental activity over the years, and led to the APS Buckley prizes of 2015.

Experimental proof for a true QCP (rather than a first order transition or a crossover in the same phase of matter) requires a diverging correlation length or timescale. In the O(N) model, such a scale is provided by the amplitude-Higgs mode, whose frequency (mass) softens as one approaches the QCP.

The Higgs mass is not protected by any symmetry, hence it is actually a resonance which decays into pairs of Goldstone modes (phasons, or spin waves, for N = 2 and N = 3, respectively, where N is the number of components of the order parameter). In three spatial dimensions (d = 3), the relative width of the Higgs resonance sharpens at the QCP. Hence, one expects the Higgs mode to be well-defined and experimentally visible all the way to the QCP.

In d = 2, however, quantum fluctuations are stronger, the Higgs does not sharpen toward the QCP, and the Higgs' visibility near criticality has been called into question. In addressing this issue, Daniel Podolsky, Dan Arovas, and I have shown that in the perturbative limit (i.e., deep in the broken symmetry phase), the Higgs mode is most visible in scalar correlation functions. These measure correlations not of the order parameter itself, nor its longitudinal component, but rather of its square. The questions which Snir Gazit's thesis set to answer were

1. Is there a well-defined Higgs peak in the scalar susceptibility close to the QCP?
2. Is there a well-defined Higgs threshold of the dynamical conductivity?

To approach the QCP, Snir used the quantum Monte Carlo "worm algorithm" of N. Prokof'ev and B. Svistunov. It allows for accurate simulations at large correlation lengths, within 2 % of the QCP. Large sample averaging was needed to reduce numerical noise and to reliably perform analytic continuation from imaginary time data to real time dynamical correlations.

This thesis presents the first calculations of the scalar susceptibility and the dynamical conductivity of the two-dimensional O(N) field theory very close to QCP. Universal spectral functions, and amplitude ratios are computed.

The results are strong and gratifying:

1. The Higgs mode in two-dimensional O(N) models remains well-defined arbitrarily close to the QCP in d = 2.
2. The Higgs mass can be measured as a peak in the scalar response function.
3. For the O(2) model (bosons), the Higgs threshold is well-defined in the dynamical conductivity.
4. Predictions for universal ratios were computed, and their values, which can be compared to experiments, are presented.
5. Charge-Vortex duality between the superfluid and the insulator ("vortex superfluid") phases has been proposed several decades ago by Matthew Fisher and Dung-Hai Lee. Using a reciprocity relation between the complex dynamical charge and vortex conductivities, this Thesis finds that the duality holds only approximately. The critical conductivity differs from q^2/h, and the amplitude ratios between are not equal on both sides of the transition.

While the main text is devoted to obtaining these results, the appendices include details of technical advances that should be useful for other computations of static and dynamical correlations in quantum systems. These include the generalization of the worm algorithm to O(N) models for N > 2, and the Singular Value Decomposition approach to analytical continuation from imaginary to real time.

Haifa, Israel
April 2015

Prof. Assa Auerbach

Abstract

Quantum phase transitions are ubiquitous in condensed matter and cold atomic systems. Some physical systems undergo a zero temperature phase transition between a disordered and a broken symmetry phase, which is tuned by a non-thermal parameter. A remarkable signature of continuous phase transitions, both in classical and in quantum systems, is the emergence of universality. In the quantum case, not only static properties are universal but also dynamical properties. In this thesis we study dynamical aspects of quantum criticality in two space dimensions. Our focus is on systems with relativistic dynamics and O(N) symmetry that is spontaneously broken in the ordered phase. To study the real-time dynamics we employ a large-scale quantum Monte Carlo simulation combined with numerical analytic continuation. We compute the universal scaling function of two experimentally pertinent response functions: the scalar susceptibility and the optical conductivity. From this analysis we deduce that the amplitude (Higgs) mode is a universal spectral feature that can be probed arbitrarily close to the critical point. Moreover, we characterize the universal properties of the amplitude mode line shape and determine the universal amplitude ratio between the amplitude mode mass and the single particle gap in the disordered phase. In addition, we study the charge-vortex duality at finite frequency near the superfluid to insulator transition. Using a generalized reciprocity relation between charge and vortex conductivities at complex frequencies, we identify the capacitance in the insulating phase as a measure of vortex condensate stiffness. We compute the ratio of boson superfluid stiffness to vortex condensate stiffness for the relativistic O(2) model. The product of dynamical conductivities at mirror points is used as a test of charge-vortex duality. Our predictions motivate future experiments that probe dynamical properties near quantum criticality.

Acknowledgments

I wish to thank my advisor, Assa Auerbach, for his guidance and support. Assa's never ending optimism was a true source of inspiration.

I am grateful to Daniel Podolsky for his close mentorship and his belief in my success. I am indebted to Daniel for teaching me how research is done.

Special thanks go to the members of the Technion condensed matter group: Emil Polturak, Amit Keren, Amit Kanigel, Nadav Shapira, Yuval Lamhot, Gil Drachuk, Tom Leviant, Ilia Khait, Omer Yair, Lee Yacobi, and Omri Bahat-Treidel. It was always fun to have a few laughs and to talk physics during the celebrated 3 PM coffee breaks.

Lastly, I wish to dedicate this work to my beloved wife Maya Epler and my parents Shlomo and Shulamit Gazit for the unconditional love and support.

The generous financial help of the Technion and the Clore Foundation are gratefully acknowledged.

Contents

Abbreviations

BHM	Bose-Hubbard Model
CVD	Charge-Vortex Duality
GLT	Ginzburg-Landau Theory
MI	Mott Insulator
QCP	Quantum Critical Point
QMC	Quantum Monte Carlo
QPT	Quantum Phase Transition
SF	Superfluid
SIT	Superfluid to Insulator Transition
SVD	Singular Value Decomposition

Chapter 1
Introduction

Some physical systems undergo a zero temperature quantum phase transition (QPT) between a disordered phase and a broken symmetry phase that is tuned by a non-thermal quantum parameter. Studying the physical properties of the aforementioned phases and tracking their evolution upon approach to the quantum critical point (QCP) is a central research theme in condensed matter and cold atomic systems.

Experimental realizations of quantum phase transitions are plentiful, among them are: the Nèel ordered to the spin gap phase in dimerized anti-ferromagnets [1, 2], the superfluid to Mott insulator transition of cold atoms trapped in an optical lattice [3], and the superconductor to insulator transition in Josephson junction arrays [4] and disordered superconductors [5].

One remarkable aspect of continuous (second order) phase transitions, both in classical and quantum systems, is the emergence of universality. It has been observed that some properties near the critical point are universal, namely they do not depend on the specific microscopic details of the system other than symmetry and dimensionality. Physical systems that share the same critical properties are said to be in the same universality class. For instance, near the critical point the correlation length ξ diverges as a power law,

$$\xi^{-1} \propto (g - g_c)^{\nu}. \tag{1.1}$$

Here g_c is the critical value of the quantum parameter g and ν is the correlation length exponent. The value of ν is universal as it depends only on the universality class.

In the quantum case, in addition to the diverging length scale there is an emergent softening energy scale Δ, which is inversely proportional to a diverging correlation time ξ_τ:

$$\Delta \propto \xi^{-z}, \tag{1.2}$$

In the above equation $z > 0$ is the dynamical critical exponent. More generally, near quantum criticality not only static properties are universal but also *dynamical* properties. Specifically, the critical energy scales associated with the collective modes and dynamical response functions such as the optical conductivity and the dynamical charge susceptibility exhibit universal behavior.

S. Gazit, *Dynamics Near Quantum Criticality in Two Space Dimensions*,
Springer Theses, DOI 10.1007/978-3-319-19354-0_1

The critical dynamical properties near the QPT are governed by the collective excitations on both sides of the phase transition. The excitation spectrum in the disordered phase consists of gapped modes. In the ordered phase, whenever the broken symmetry is continuous, there are two types of collective excitations: Goldstone modes, which describe fluctuations in the transverse direction with respect to the broken symmetry direction and a single gapped amplitude (Higgs) mode that corresponds to fluctuations of the order parameter amplitude.

Duality transformations, in statistical mechanics models, map the high temperature region onto the low temperature region. In the language of quantum field theory the mapping is between small and large coupling constants. Examples of duality mappings are: the Kramers-Wannier duality of the Ising model and the Coulomb gas description of the XY model in two dimensions. The degrees of freedom of the dual theory are "disorder variables" such as spin flips in the Ising model and vortices in the XY model. The dual description provides further insight into the problem and in certain cases it yields exact results.

The research thesis consists of two projects, in which we study some of the foregoing aspects of dynamics near quantum criticality:

- **Dynamics and conductivity near quantum criticality in two space dimensions**

  While the Goldstone modes are stable modes, protected by symmetry, the amplitude mode can decay into a pair of Goldstone modes. This broadens the spectral line and might reduce the visibility of the amplitude mode in real experiments. The situation is particularly interesting in two space dimensions, where quantum corrections are strong and thus can render the amplitude resonance over-damped. In this research thesis we demonstrate that the amplitude mode is a well defined spectral feature arbitrarily close to the critical point in the spectrum of two experimentally relevant response functions: the scalar susceptibility and the optical conductivity. In addition, we characterize the universal proprieties of the amplitude mode. The results are pertinent to recent experimental studies of critical dynamics in cold atomic systems and granular superconductors.

- **Critical capacitance and charge-vortex duality near the superfluid to insulator transition**

  The superfluid to insulator transition (SIT) in two space dimensions has an interesting dual description in terms of vortex degrees of freedom. In this language the insulator is a *Bose condensate* of vortices. The duality mapping is not self dual, mainly due to the different interaction ranges of bosons and vortices. The problem with this approach is that it lacks an experimental probe for the vortex condensate. In our study, we identify the capacitance in the insulating phase as a direct measure of vortex condensate stiffness. In addition, we show that product of the optical conductivity evaluated at mirror points across the phase transition can be used as quantitative measure for the deviation from self duality.

In the following we present a short introduction to the research projects. A detailed description of the research methods and results is given in Chaps. 2 and 3 and in the appendices.

1.1 Spontaneous Symmetry Breaking

In many systems of interest, the effective long wavelength field theory near the qunatum critical point (QCP) is captured by a ϕ^4 Ginzburg-Landau theory (GLT):

$$\mathcal{Z} = \int \mathcal{D}\phi_\alpha e^{-\mathcal{S}[\phi_\alpha]} \tag{1.3}$$
$$\mathcal{S} = \int d^D x \left[(\partial_\mu \phi_\alpha)^2 + \mu\phi_\alpha^2 + g(\phi_\alpha^2)^2 \right].$$

Here $\mathcal{Z}$ is the partition function defined on a $D = d+1$ dimensional Euclidean space time $\vec{x} = \{t, \vec{r}\}$, $\partial_\mu = \{\partial_\tau, \partial_r\}$ and ϕ_α is an N component vector order parameter. The action is $O(N)$ symmetric and the dynamics is Lorentz invariant.

Relativistic dynamics, $z = 1$, is natural in high energy physics whereas in condensed matter and cold atomic systems this property is typically an emergent symmetry. Some examples of such systems are: dimerized antiferomagnets for $N = 3$ and the Bose-Hubbard model at *integer* filling for $N = 2$.

The mean field phase diagram of Eq. (1.3) is derived by neglecting fluctuations of the order parameter. Explicitly, the mean field order parameter is taken to be uniform, $\langle \phi_\alpha(x) \rangle = \bar{\phi}_\alpha$, and its value is determined by minimizing the mean field effective potential:

$$V_{MF}(\bar{\phi}) = \mu\bar{\phi}^2 + g(\bar{\phi}^2)^2 \tag{1.4}$$

This analysis predicts a quantum phase transition between a disordered and an ordered phase. In the disordered phase, $\mu > 0$, the effective potential of the order parameter, depicted in Fig. 1.1a, has a single minimum, where the expectation value of the order parameter vanishes $\bar{\phi} = 0$. In the ordered phase, $\mu < 0$, the effective potential, as shown in Fig. 1.1b, has a continuous set of minima along the rim of the "Mexican Hat" potential with a non vanishing expectation value $\bar{\phi} > 0$. The ground state spontaneously breaks the $O(N)$ symmetry by picking one specific minimum.

The mean field approximation must be corrected by taking into account the effect of fluctuations. The stability of the mean field result with respect to fluctuations depends on dimensionality: Above the upper critical dimension, $D > 4$, the fluctuations are suppressed and the corrections are only quantitative. By contract, below the lower critical dimension $D < 2$ the order is completely destroyed due to strong fluctuations.

In this thesis we will focus mainly on two dimensional quantum systems ($D = 3$). Here, the mean field predictions are valid only far from the phase transition and the fluctuations have prominent effect near the critical point. The crossover between

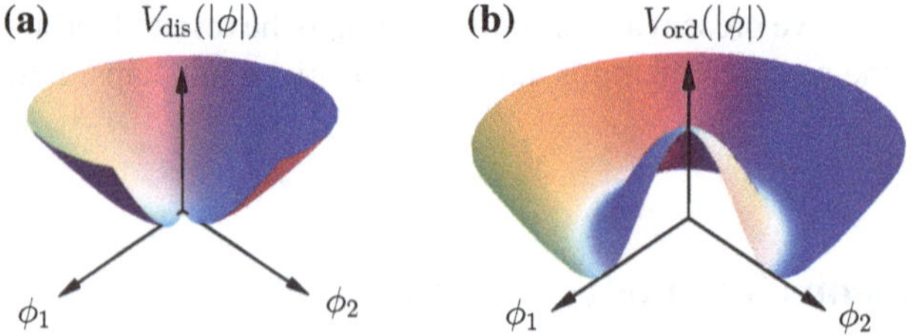

Fig. 1.1 The mean field effective potential for $N = 2$ in Eq. (1.4). **a** In the disordered phase the effective potential has a single minimum where $\langle\vec{\phi}\rangle = 0$. **b** In the ordered phase there is a continuous set of minima along the rim of the "Mexican Hat" potential, where the order parameter has a non zero expectation value $\langle\vec{\phi}\rangle \neq 0$

the two limits is set by the Ginzburg criteria defined as the ratio between the order parameter amplitude and the standard deviation of the fluctuations. Importantly, the *critical* properties differ significantly from the mean field predictions. For instance, the numerical estimate of the correlation length critical exponent for $N = 2$ is $\nu \approx 0.67$, whereas the mean field prediction is $\nu = 1/2$.

1.2 Superfluid to Mott Insulator Transition of Lattice Bosons at Integer Filling

The Bose Hubbard model (BHM) is a concrete example for a condensed matter model where relativistic dynamics is realized as an emergent symmetry.

The BHM describes a system of interacting lattice bosons. The Hilbert space is a Fock space of bosonic occupation numbers, $\{|n_i\rangle\}$, at each site. We define $b_i^\dagger$ (b_i) as the bosonic creation (annihilation) operators at the site i, with the canonical commutation relations $\left[b_i, b_i^\dagger\right] = \delta_{i,j}$. The boson number operator at site i is then given by $n_i = b_i^\dagger b_i$. The Hamiltonian of the BHM is defined as:

$$\mathcal{H} = -J \sum_{\langle i,j\rangle} b_i^\dagger b_j + b_j^\dagger b_i - \mu \sum_i n_i + U \sum_i n_i(n_i - 1) \tag{1.5}$$

The first term corresponds to the kinetic energy of the bosons, describing hopping processes between adjacent sites. The second term sets the average particle number by tuning the chemical potential μ. The last term is an onsite repulsive interaction term of strength U. Particle number is a conserved quantity since the Hamiltonian commutes with the total number operator $N = \sum_i n_i$. Equivalently, the model has a global $U(1)$ symmetry under the transformation $b_i \to b_i e^{i\theta}$.

The competition between the kinetic energy term and the interaction term results in a zero temperature quantum phase transition between a superfluid (SF) and a Mott insulator (MI).

The MI phase is best understood in the atomic limit, $J/U \to 0$. The ground state is then a product state, with exactly n bosons at each site:

$$|\psi\rangle \propto \prod_i (b_i^\dagger)^n |0\rangle \tag{1.6}$$

The MI is gapped since adding an extra boson has an energy cost U. Therefore, the state is incompressible, i.e. $\partial n/\partial \mu = 0$. Tuning the chemical potential μ leads to a sequence of Mott states with integer fillings, $n = 1, 2, 3, \ldots$.

At finite $J/U > 0$ the hopping term delocalizes the bosons and further increase of J/U drives a QPT, at a critical coupling $J_c(U)$, between the MI phase and the SF phase. In the SF phase the bosons condense and the $U(1)$ symmetry is spontaneously broken. Deep inside the superfluid phase, $J/U \gg 1$, the ground state wave function is approximately a zero momentum condensate:

$$|\psi\rangle \approx \left(\sum_i b_i^\dagger\right)^N |0\rangle \tag{1.7}$$

The full phase diagram as a function of J/U and μ/J is summarized pictorially in Fig. 1.2

To capture the critical properties near the QPT and to relate them to GLT in Eq. (1.3) we write an effective GLT action for the phase transition in terms of a complex order parameter Ψ preserving the $U(1)$ symmetry of the model:

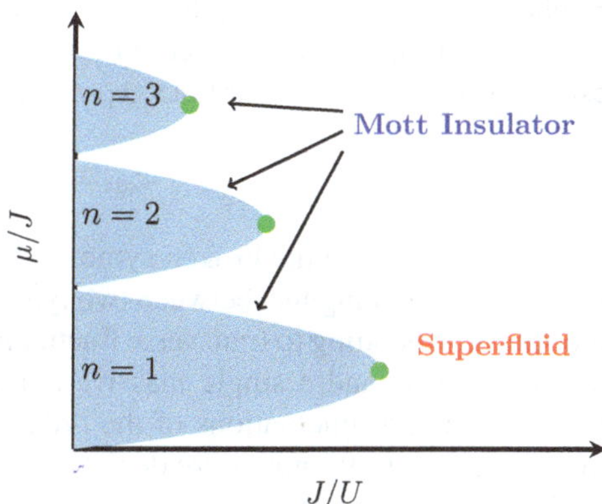

Fig. 1.2 Phase diagram of the Bose-Hubbard model in the J/U versus μ/J plane. The *green* circles at the tip of the Mott lobes corresponds to the multicritical points, where the dynamics is relativistic

$$\mathcal{S}_B = \int d\tau \, d^d x \left[K_0 \Psi^* \frac{\partial \Psi}{\partial \tau} + K_1 \left| \frac{\partial \Psi}{\partial \tau} \right|^2 + K_2 \left| \nabla \Psi \right|^2 \right.$$
$$\left. + r \left| \Psi \right|^2 + u \left| \Psi \right|^4 \right] \tag{1.8}$$

The GLT can be derived directly from the microscopic model of Eq. (1.5), by applying a Hubbard-Stratonovitch transformation [6] and expanding the action in the limit of small order parameter and long wavelength. The linear time derivative term does not break any symmetry and hence can not be excluded from the effective action.

The coefficient of the linear derivative term K_0 is related to the coefficient of the quadratic term, r, in Eq. (1.8) and hence it is not independent [7]:

$$K_0 = -\frac{\partial r}{\partial \mu}. \tag{1.9}$$

In order for K_0 to vanish we must have that $\frac{\partial r}{\partial \mu} = 0$. This condition is satisfied exactly at the tip of the Mott lobe, marked by green dots in Fig. 1.2, where the phase boundary between the SF and the MI is vertical. Hence, the resulting theory has an emergent Lorentz symmetry, and it is equivalent to the relativistic QPT model presented in Eq. (1.3) with $N = 2$. The two components of the vector field $\vec{\phi}$ are identified with the real and imaginary part of the complex order parameter Ψ.

1.3 Collective Excitations

The excitations spectrum, beyond the mean field ground state, is obtained by considering small fluctuations of the order parameter above the mean field solution. Explicitly, we take $\phi_\alpha = \bar{\phi}_\alpha + \delta\phi_\alpha$ and expand the action to quadratic order in $\delta\phi_\alpha$.

The resulting excitation spectrum in the disordered phase consists of N massive excitations with a single particle gap Δ and the dispersion relation:

$$\omega^2_{\text{dis}}(k) = k^2 + \Delta^2 \tag{1.10}$$

In the ordered phase the broken symmetry direction is a preferred direction. Therefore, we must distinguish between two types of excitations: $N-1$ gapless Goldstone modes corresponding to transverse fluctuations, $\delta\phi_t$, with respect to the broken symmetry direction and a single massive amplitude (Higgs) mode, with a mass m_H, corresponding to fluctuations of the order parameter amplitude in the longitudinal direction, $\delta\phi_\ell$. Both modes are depicted pictorially in Fig. 1.3a.

The dispersion relation of the two modes is shown in Fig. 1.3b and is given by:

$$\omega^2_{\text{t}}(k) = k^2 \tag{1.11}$$
$$\omega^2_{\ell}(k) = k^2 + m^2_H,$$

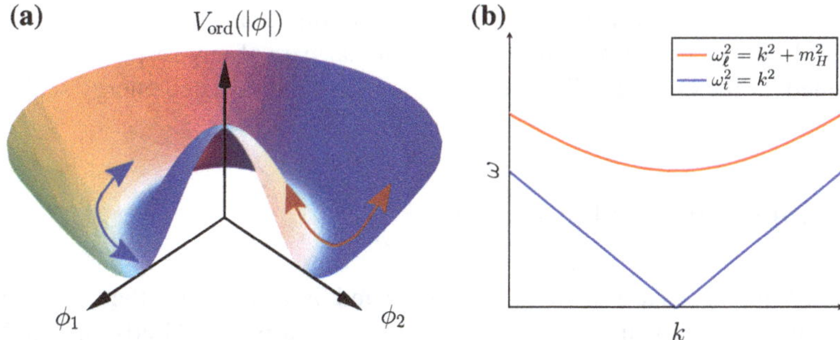

Fig. 1.3 Excitation spectrum in the ordered phase. **a** The Golstone mode (*blue*) corresponds to longitudinal fluctuations, whereas the amplitude mode (*red*) describe transverse fluctuations of the order parameter amplitude. **b** Dispersion relations of the massless Goldstone mode and the gapped amplitude mode

Here m_H is the mass of amplitude mode.

The mean field analysis predicts that both energy scales soften to zero towards the critical point, following a singular power law behavior:

$$\Delta, m_H \propto (g - g_c)^\nu \tag{1.12}$$

The softening energy scales signals the divergence of the correlation time.

1.4 Universality and the Scaling Limit

One of the defining properties of second order phase transitions is the divergence of the correlation length ξ and correlation time ξ_τ near the critical point. In cases where the dynamics is relativistic, space and time scales are equivalent up to rescaling by the speed of sound. As a result, the low energy physics is characterized solely by the vanishing critical energy scale $\Delta \sim \xi_\tau^{-1}$.

The scaling limit is then obtained by considering the theory in the limit where $\xi/a \gg 1$, with a being the microscopic lattice constant scale. The assertion of universality is that the result of the scaling limit is independent of the microscopic details of the system other than symmetries and dimensionality.

One important result of the universality assumption is that the long wave length critical properties of any correlation function, $\chi(\omega)$, can be expressed in terms of a universal scaling function:

$$\chi(\omega, g) \propto \Delta^{d_\chi} f(\omega/\Delta). \tag{1.13}$$

Here d_χ is the scaling dimension of the response function $\chi(\omega)$ and $f(x)$ is a universal scaling function. Importantly, the universal part of any response function is determined by the critical collective excitations and their related energy scales.

1.5 Visibility of the Higgs Mode

The Goldstone modes are stable modes, protected by Goldstone's theorem, whereas the amplitude mode is unstable as it can decay into a pair of Goldstone modes by quantum corrections that decrease the life time and hence broaden the experimental line shape.

A natural response function for probing the amplitude mode is the *longitudinal* susceptibility:

$$\chi_\ell(q) = \int d^D x e^{iq\cdot x} \langle \delta\phi_\ell(x)\delta\phi_\ell(0)\rangle \tag{1.14}$$

In the above equation $\vec{q} = \{\omega, \vec{p}\}$ is the energy-momentum vector. The visibility of the amplitude mode in the spectrum of the longitudinal susceptibility depends on dimensionality. In $D = 3+1$ dimensions the amplitude mode becomes increasingly sharper upon approach to the critical point [8]. By contrast, in $D = 2+1$ dimensions the longitudinal susceptibility has a $1/\omega$ divergence at low frequency such that the amplitude mode is concealed in that limit [9].

Recently, it was understood that the line shape of the amplitude mode is sensitive to the symmetry of the probe [10]. It was suggested to consider the *scalar* susceptibility in order to improve the visibility of the amplitude mode. The scalar susceptibility measures the response to an experimental probes that couple to the order parameter amplitude *squared*:

$$\chi_s(q) = \int d^D x e^{iq\cdot x} \langle \phi^2(x)\phi^2(0)\rangle \tag{1.15}$$

The scalar susceptibility was computed in the large N and weak coupling limits [10]. It was found that, unlike the longitudinal susceptibility, the scalar susceptibility rises as ω^3 at low frequency, rendering the amplitude mode visible at finite frequencies. A comparison between the longitudinal and scalar susceptibilities, computed in the large N limit, is shown in Fig. 1.4. The amplitude mode is visible in the spectrum of the scalar susceptibility, whereas the line shape is completely washed out in the longitudinal susceptibility.

The weak coupling expansion and the large N limit are valid only deep inside the superfluid phase. This is due to the fact that the critical point, in $D = 2+1$, is a strongly coupled fixed point [11], precluding a description in terms of weekly interacting normal modes. As a consequence, these methods might not capture correctly the universal properties near the critical point. In particular, it not obvious that the amplitude mode survives as a well defined resonance arbitrary close to the critical point.

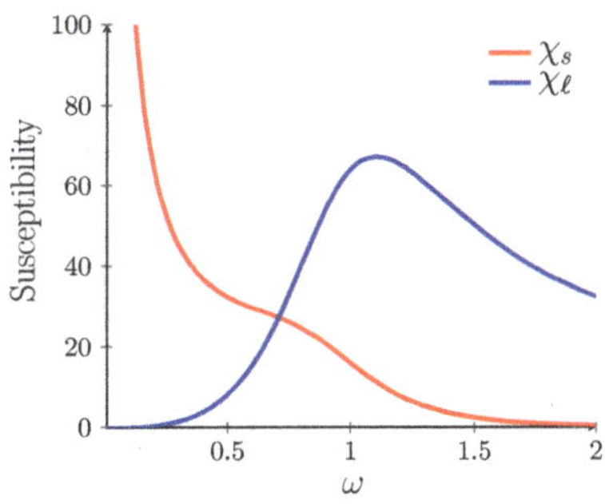

Fig. 1.4 A comparison between the scalar susceptibility (*blue*) as defined in Eq. (1.15) and the longitudinal susceptibility (*red*) as defined in Eq. (1.14). Both curve were computed in the large N limit

To compute the universal scaling function of the scalar susceptibility the authors of Ref. [12] considered the leading order correction to the large N limit in a $1/N$ expansion. The conclusion of this analysis was that the universal part of the scalar susceptibility contains a well defined resonance that can be associated with the amplitude mode.

The $1/N$ expansion is an asymptomatic expansion, which is well controlled only in the large N limit. Hence, these results might be inaccurate for the experimentally relevant cases, where N is small ($N = 2$ for the Bose Hubbard model and $N = 3$ for dimerized antiferromagnets). Therefore, determining the ultimate fate of the amplitude mode near quantum criticality requires an unbiased numerical computation. In Chap. 2 we present a large scale quantum Monte Carlo (QMC) simulation supplemented by a numerical analytic continuation, where we study the visibility of the amplitude mode in the spectrum of the scalar susceptibility and the optical conductivity near quantum criticality.

1.6 Experimental Motivation: The Amplitude (Higgs) Mode Near the Two Dimensional Superfluid to Mott Insulator Transition

Cold atom systems provide highly tunable experiments of strongly correlated quantum systems [3, 13]. One prominent example is the superfluid to Mott insulator transition of cold atoms trapped in an optical lattice. This system is well described by the BHM presented in Eq. (1.5). The hopping term corresponds to tunneling events between adjacent sites of the optical lattice. The hopping rate is tuned by varying the depth of the optical potential. The Hubbard U interaction term models on-site repulsive interaction between the atoms [13].

The first experimental observation of the amplitude (Higgs) mode near quantum criticality in two space dimensions was in recent breakthrough experiment of the SF to MI transition at integer filling [14].

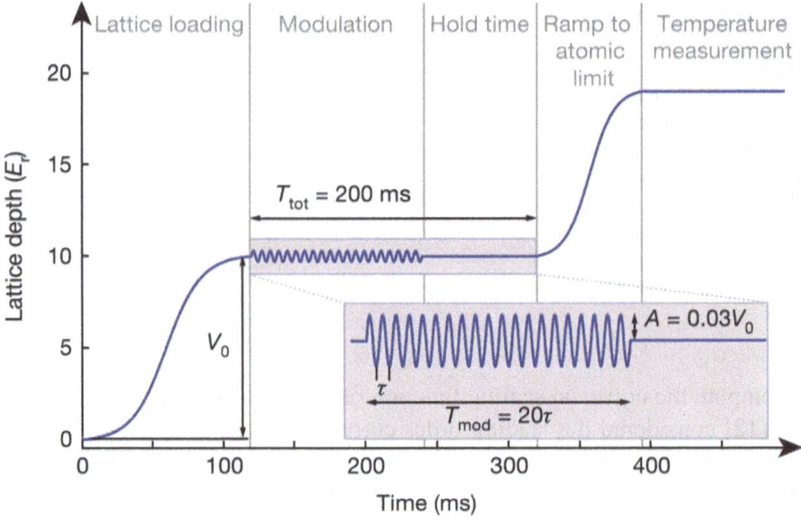

Fig. 1.5 Experimental protocol for the detection of the amplitude mode near the superfluid to Mott insulator transition in cold atoms trapped in an optical lattice. The system is perturbed by modulating the lattice depth and the spectral response function is obtained by measuring the temperature increase by means of single site imaging techniques

The experimental protocol is depicted schematically in Fig. 1.5. First, a degenerate gas of ^{87}Rb atoms is loaded into a two dimensional optical lattice generated by a standing wave pattern of laser light. The detuning from the QCP is determined by the lattice depth. Next, the system is perturbed by weak modulations of the lattice depth followed by a hold time which allows equilibration. Finally, the energy absorption is measured by increasing the potential depth to reach the atomic limit and measuring the temperature increase by single atom resolving techniques [15].

The experimentally measured response function in this case is given by the dynamical kinetic energy correlation function [16]. In order to relate this response function to the scalar susceptibility, we investigate the effective action near the quantum critical point in Eq. (1.8). The parameters K_1, K_2, r, u are all functions of the hopping amplitude J and the on-site interaction U. For instance, if the transition is tuned by varying the hopping amplitude J then r can be linearized about the critical point as $r = r_0(J - J_c)$. Thus, the energy absorbed due to modulations of the hopping amplitude J can be described by the dynamical response of an operator of the form:

$$\Theta(x, \tau) = \alpha|\Psi|^2 + \beta|\partial_\tau \Psi|^2 + \gamma|\nabla\Psi|^2 + \delta|\Psi|^4 \tag{1.16}$$

Close to the QCP, the term proportional to $|\Psi|^2$ will dominate the low energy spectral weight, since it is the most relevant operator in the long wavelength limit. As a consequence, the universal part of the response function coincides with the scalar susceptibility.

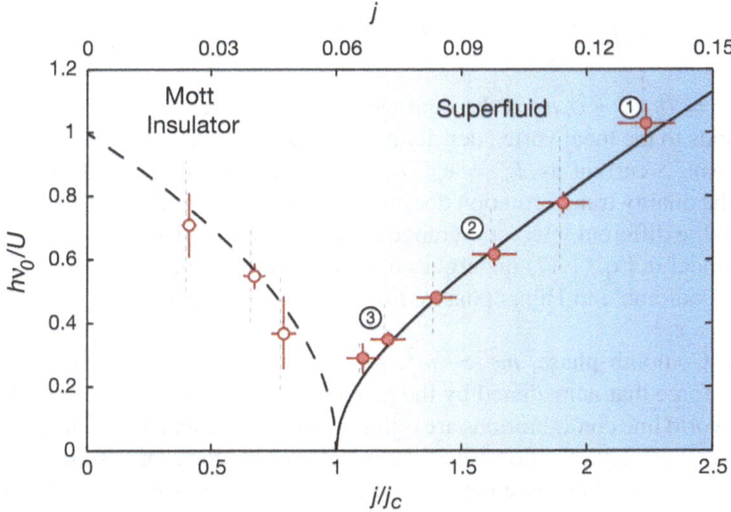

Fig. 1.6 Critical energy scales measured from the spectral response function to lattice depth modulations. The softening energy scales correspond to twice the charge gap in the Mott phase and to the amplitude (Higgs) resonance in the superfluid phase

The experimentaly resolved energy scales near the critical point are shown in Fig. 1.6. The softening energy scales correspond to twice the charge gap in the Mott phase and to the amplitude (Higgs) resonance in the superfluid phase.

Unfortunately, due to experimental limitations, the detuning from the QCP was not small enough to probe the critical region. This motivates future experiments that study the universal properties of the amplitude mode in close proximity to the QCP.

1.7 Charge Vortex Duality Near the Superfluid to Insulator Transition

The duality mapping between bosons and vortices plays an important role in various strongly interacting systems [17]. A few examples are: the Coulomb gas description of the two dimensional XY model [18] and fractional charges in fractional quantum Hall systems [19].

In this section we will discuss the dual vortex description of interacting bosons in two dimensions [20]. The focus will be on the particle-hole symmetric case where the dynamics is relativistic, i.e. K_0 vanishes in Eq. (1.8). The effective action of the dual vortex representation is a complex $|\varphi|^4$ theory minimally coupled to an emergent abelian gauge field a_μ [20–22]:

$$\mathcal{S}_{\rm V} = \int d^2x \int d\tau \left[\left|(\partial_\mu - i a_\mu)\varphi\right|^2 + m^2|\varphi|^2 + u|\varphi|^4 + \frac{1}{4e^2} f_{\mu\nu}^2 \right]. \tag{1.17}$$

Here $f_{\mu\nu} = \partial_\mu a_\nu - \partial_\nu a_\mu$ is the dual electromagnetic tensor, the complex field φ corresponds to the local vortex density and the dual abelian gauge field a_μ is related to the boson 3-current as $J_\mu = \epsilon_{\mu\nu\lambda}\partial_\nu a_\lambda$. The duality mapping is not self dual, namely the duality transformation does not map the model to itself. This is attributed mainly to the different interaction range of bosons and vortices.

The model in Eq. (1.17) undergoes a zero temperature quantum phase transition between Coulomb and Higgs phases, tuned by the mass parameter m^2 at a critical value $m^2 = m_c^2$.

In the Coulomb phase, $m^2 > m_c^2$, the vortices interact through a long range Coulomb force that is mediated by the gauge field a_μ. Pictorially, in this phase, the vortices world line configurations are tightly bound vortex anti-vortex loops confined by the strong attractive Coulomb interaction. This phase corresponds to the superfluid phase of the original bosonic theory. The gapless phase mode (Goldstone mode) is identified with the gapless transverse mode of the gauge field a_μ.

In the Higgs phase, $m^2 > m_c^2$, the vortex field φ condenses. The non zero expectation value, $\bar{\varphi} > 0$, of the vortex field serves as a mass term for the gauge field a_μ through the Higgs-Anderson mechanism [23]. As a consequence, the interaction between the vortices is screened, resulting in a proliferation of large vortex anti-vortex loops. Due to the mass term, the gauge field a_μ supports an additional a longitudinal mode thus it is no longer purely transverse. This phase is essentially a vortex Bose condensate that corresponds to the insulating phase of the original bosonic model. The two gapped modes (particle and hole excitations) of the insulator are identified with the two gapped modes of the gauge field a_μ.

Interestingly, the conductivity of the bosons, σ, is inversely proportional to the conductivity of the vortices, $\sigma_{\rm V}$:

$$\sigma \times \sigma_{\rm V} = \frac{q^2}{\hbar^2}. \tag{1.18}$$

Here q is the charge of the bosons ($= 2e$ in superconductors). This exact relation is fully consistent with the physical picture of the dual vortex model. The bosonic superfluid, with $\sigma = \infty$, is mapped to a vortex insulator, with $\sigma_{\rm V} = 0$. Similarly, the bosonic insulator, with $\sigma = 0$, is mapped to a vortex condensate, with $\sigma_{\rm V} = \infty$. It was further speculated [24] that if the duality is approximately self dual, i.e. $\sigma_{\rm V} = \sigma/q^2$, then the conductivity at the critical point equals exactly to the quantum of conductance $\sigma_q = q^2/h$. A value that is close to the experimental observation [5].

The problem with the vortex description is that it lacks a concrete observable, which probes the vortex condensate in terms of the physical bosonic degrees of freedom. The purpose of the second project, presented in Chap. 3, is to address this issue. We use the reciprocity relation in Eq. (1.18) to identify the capacitance in the insulating phase as a direct probe for the dual vortex condensate stiffness. In addition, we show that the universal ratio between the superfluid stiffness and the dual vortex condensate stiffness, evaluated at mirror points across the phase transition, is a quantitative measure for the deviation from self-duality.

References

1. S. Chakravarty, B.I. Halperin, D.R. Nelson, Two-dimensional quantum Heisenberg antiferromagnet at low temperatures. Phys. Rev. B **39**(4), 2344–2371 (1989). doi:10.1103/PhysRevB.39.2344. http://link.aps.org/doi/10.1103/PhysRevB.39.2344. Accessed 22 Oct 2012
2. CH. Ruegg et al., Quantum magnets under pressure: controlling elementary excitations in TlCuCl 3. Phys. Rev. Lett. **100**(20), 205701 (2008). doi:10.1103/PhysRevLett.100.205701. http://link.aps.org/doi/10.1103/PhysRevLett.100.205701. Accessed 22 Oct 2012
3. M. Greiner et al., Quantum phase transition from a super fluid to a Mott insulator in a gas of ultracold atoms. Nature 415.6867, 39–44. ISSN 0028-0836 (2002). doi:10.1038/415039a. http://dx.doi.org/10.1038/415039a
4. L.J. Geerligs et al., Charging effects and quantum coherence in regular Josephson junction arrays. Phys. Rev. Lett. **63**(3), 326–329 (1989). doi:10.1103/PhysRevLett.63.326. http://link.aps.org/doi/10.1103/PhysRevLett.63.326
5. A.M. Goldman, Superconductor-insulator transitions. Int. J. Mod. Phys. B 24.20n21, 4081–4101 (2010). doi:10.1142/S0217979210056451. http://www.worldscientific.com/doi/abs/10.1142/S0217979210056451
6. K. Sengupta, N. Dupuis, Mott-insulator to super fluid transition in the Bose-Hubbard model: a strong-coupling approach. Phys. Rev. A **71**(3), 033629 (2005). doi:10.1103/PhysRevA.71.033629. http://link.aps.org/doi/10.1103/PhysRevA.71.033629
7. M.P.A. Fisher et al, Boson localization and the super fluid-insulator transition. Phys. Rev. B **40**(1), 546–570 (1989). doi:10.1103/PhysRevB.40.546. http://link.aps.org/doi/10.1103/PhysRevB.40.546. Accessed 23 Oct 2012 (73 Bibliography 74)
8. I. Aeck, G.F. Wellman, Longitudinal modes in quasi-one-dimensional antiferromagnets. Phys. Rev. B **46**(14), 8934–8953 (1992). doi:10.1103/PhysRevB.46.8934. http://link.aps.org/doi/10.1103/PhysRevB.46.8934. Accessed 22 Oct 2012
9. N. Dupuis, Infrared behavior in systems with a broken continuous symmetry: classical O(N) model versus interacting bosons. Phys. Rev. E **83**(3), 031120 (2011). doi:10.1103/PhysRevE.83.031120. http://link.aps.org/doi/10.1103/PhysRevE.83.031120
10. D. Podolsky, A. Auerbach, D.P. Arovas, Visibility of the amplitude (Higgs) mode in condensed matter. Phys. Rev. B **84**(17), 174522 (2011). doi:10.1103/PhysRevB.84.174522. http://link.aps.org/doi/10.1103/PhysRevB.84.174522
11. Kenneth G. Wilson, M.E. Fisher, Critical exponents in 3.99 dimensions. Phys. Rev. Lett. **28**(4), 240–243 (1972). doi:10.1103/PhysRevLett.28.240. http://link.aps.org/doi/10.1103/PhysRevLett.28.240
12. D. Podolsky, S. Sachdev, Spectral functions of the Higgs mode near two-dimensional quantum critical points. Phys. Rev. B **86**(5), 054508 (2012). doi:10.1103/PhysRevB.86.054508. http://link.aps.org/doi/10.1103/PhysRevB.86.054508
13. I. Bloch, J. Dalibard, W. Zwerger, Many-body physics with ultracold gases. Rev. Mod. Phys. **80**(3), 885–964 (2008). doi:10.1103/RevModPhys.80.885. http://link.aps.org/doi/10.1103/RevModPhys.80.885
14. M. Endres et al, The /'Higgs/' amplitude mode at the two-dimensional super fluid/Mott insulator transition. en. Nature **487**(7408), 454–458 (2012). ISSN 0028-0836, doi:10.1038/nature11255. http://www.nature.com/nature/journal/v487/n7408/full/nature11255.html Accessed 11 Oct 2012
15. J.F. Sherson et al, Single-atom-resolved fluorescence imaging of an atomic Mott insulator. Nature 467.7311, 68–72 (2010). ISSN 0028-0836, doi:10.1038/nature09378

16. L. Pollet, N. Prokof'ev, Higgs mode in a two-dimensional super fluid. Phys. Rev. Lett. **109**(1), 010401 (2012). doi:10.1103/PhysRevLett.109.010401. http://link.aps.org/doi/10.1103/PhysRevLett.109.010401
17. R. Savit, Duality in field theory and statistical systems. Rev. Mod. Phys. **52**(2), 453–487 (1980). doi:10.1103/RevModPhys.52.453. http://link.aps.org/doi/10.1103/RevModPhys.52.453. (Bibliography 75)
18. J.M. Kosterlitz, D.J. Thouless, Ordering, metastability and phase transitions in two-dimensional systems. J. Phys. C: Solid State Phys. **6**(7), 1181 (1973). http://stacks.iop.org/0022-3719/6/i=7/a=010
19. D.-H. Lee, M.P.A. Fisher, Anyon superconductivity and the fractional quantum Hall effect. Phys. Rev. Lett. **63**(8), 903–906 (1989). doi:10.1103/PhysRevLett.63.903. http://link.aps.org/doi/10.1103/PhysRevLett.63.903
20. M.P.A. Fisher, D.H. Lee, Correspondence between two-dimensional bosons and a bulk superconductor in a magnetic field. Phys. Rev. B **39**(4), 2756–2759 (1989). doi:10.1103/PhysRevB.39.2756. http://link.aps.org/doi/10.1103/PhysRevB.39.2756
21. M.E. Peskin, Mandelstam-'t Hooft duality in abelian lattice models. Ann. Phys. **113**(1), 122–152 (1978). ISSN 0003-4916. doi:10.1016/0003-4916(78)90252-X. http://www.sciencedirect.com/science/article/pii/000349167890252X
22. C. Dasgupta, B.I. Halperin, Phase transition in a lattice model of superconductivity. Phys. Rev. Lett. **47**(21), 1556–1560 (1981). doi:10.1103/PhysRevLett.47.1556. http://link.aps.org/doi/10.1103/PhysRevLett.47.1556
23. P.W. Anderson, Plasmons, gauge invariance, and mass. Phys. Rev. **130**(1), 439–442 (1963). doi:10.1103/PhysRev.130.439. http://link.aps.org/doi/10.1103/PhysRev.130.439
24. M.P.A. Fisher, G. Grinstein, S.M. Girvin, Presence of quantum diffusion in two dimensions: Universal resistance at the superconductor-insulator transition. Phys. Rev. Lett. **64**(5), 587–590 (1990). doi:10.1103/PhysRevLett.64.587. http://link.aps.org/doi/10.1103/PhysRevLett.64.587

Chapter 2
Dynamics and Conductivity Near Quantum Criticality

2.1 Introduction

In this chapter we study the dynamical properties of relativistic O(N) models close to the quantum critical point at low temperature, frequency, and zero wave vector. We compute the universal line shape of the scalar susceptibility for O(N) models with $N = 2, 3$, and 4. In addition, we perform a careful analysis of the low frequency behavior of the line shape in the ordered phase, where we confirm the ω^3 rise for $N = 3$ and $N = 4$ predicted in Ref. [1]. For $N = 2$ we cannot resolve the low frequency power law. The scalar response in the disordered phase exhibits a sharp threshold above a gap.

We present QMC and analytic results for the dynamical conductivity of the O(2) model on both sides of the transition. In the superfluid phase, we find a threshold-like behavior in the conductivity, which rises quadratically with frequency above the Higgs mass m_H. In the insulator there is a low-frequency threshold in the conductivity appearing at twice the single particle gap Δ, and a negative (capacitive) linear dependence of the imaginary conductivity.

Throughout the analysis we identify a number of universal constants that characterize the critical point. These include ratios of quantities measured on mirror points on the ordered/disordered sides of the transition, such as m_H/Δ and Υ/Δ, where Υ is the helicity modulus in the ordered phase (superfluid stiffness in the superfluid phase). For $N = 2$, we compute the high frequency universal conductivity $\sigma_c^*(\omega \gg T)$ in the quantum critical regime.

Our results are relevant to recent experiments which probe critical dynamics. In cold atomic gases, the Higgs mode has been excited by modulating the lattice potential near the superfluid to Mott transition [2]. Fast real time pump-probe response was used to see amplitude oscillations in charge density wave (CDW) systems [3, 4]. Raman and neutron scattering have long identified a "two magnon peak" in antiferromagnets [5–9]. Within our theory, this peak is a Higgs mode which would soften at criticality. The conductivity in cold atom systems may be measured by lattice phase modulations [10]. For Josephson junction arrays and granular

S. Gazit, *Dynamics Near Quantum Criticality in Two Space Dimensions*,
Springer Theses, DOI 10.1007/978-3-319-19354-0_2

superconducting films, Coulomb interactions must be considered, as they give rise to massless two-dimensional plasmons. We show that this increases the power law rise of the conductivity above the Higgs threshold. While our theory is for translationally invariant systems, some of the finite frequency zero wave vector results may be a good starting point for understanding very recent results on disordered granular superconducting films [11].

This chapter is organized as follows. Section 2.2 presents the $O(N)$ field theory and the observables we study, together with their expected scaling near the quantum critical point. Section 2.3 introduces the discretized lattice model. In Sect. 2.4, we locate the critical point as a function of cutoff parameters and compute the relevant energy scales near the critical point. In Sect. 2.5, we present the universal scaling functions of the scalar susceptibility. In Sect. 2.6, we compute the dynamical conductivity on both sides of the superfluid-Mott transition Appendix A describes the QMC algorithm in detail. Appendix B discusses the numerical analytical continuation procedure and provides an error analysis of the kernel pseudo-inversion. Finally, Appendix C describes a weak coupling analytic calculation of the conductivity.

2.2 Field Theory and Scaling

We consider microscopic systems with $O(N)$ symmetry whose long wave length and low energy universal properties near the QCP are captured by a quartic field theory with relativistic dynamics as presented in Eq. (1.3).

We study two dynamical observables: the scalar susceptibility and the dynamical conductivity. For completeness we define these observables and discuss their expected scaling behavior and experimental realizations.

2.2.1 Scalar Susceptibility

The scalar susceptibility describes the response function of experimental probes that are sensitive to the amplitude of the order parameter, but not to its direction [1]. The scalar susceptibility as a function of Matsubara frequency is defined similarly to Eq. (1.15) as the correlation function of the order parameter amplitude squared:

$$\begin{aligned} \chi_s(\tau) &= \int d^2x \left(\langle \vec{\phi}^{\,2}_{x,y,\tau} \vec{\phi}^{\,2}_{\mathbf{0}} \rangle - \langle \vec{\phi}^{\,2}_{\mathbf{0}} \rangle^2 \right) \\ \chi_s(i\omega_m) &= \int_0^\beta d\tau e^{i\omega_m \tau} \chi_s(\tau) \end{aligned} \tag{2.1}$$

The real frequency spectral function is obtained by analytic continuation of Eq. (2.1)

$$\chi_s''(\omega) = -Im\chi_s(i\omega_m \to \omega + i0^+) \tag{2.2}$$

Scaling arguments indicate that the expected low energy form of Eq. (2.1) near the QCP is [12]:

$$\chi_s(\omega/\Delta) \sim C + \mathcal{A}_\pm \Delta^{3-2/\nu} \Phi_\pm(\omega/\Delta) \tag{2.3}$$

where $\Delta \sim |\delta g|^\nu$ is the gap in the disordered phase, ν is the correlation length critical exponent, and Φ_- (Φ_+) is a universal function of ω/Δ on the ordered (disordered) side of the transition. The non-universal constant C is real, and is a regular function of g across the transition. The ordered phase is gapless due to the presence of Goldstone modes. In order to provide a well-defined energy scale that characterizes fluctuations on the ordered phase ($\delta g < 0$), we use the gap at the mirror point $-\delta g$ across the transition.

2.2.2 Conductivity

The dynamical conductivity measures the response to an external gauge field. Our analysis will be restricted to the $N = 2$ case, as is relevant to dynamical conductivity measurements in superconductors and also to neutral cold atoms probed by optical lattice phase modulations [10]. To simplify the analysis we write the two scalar fields in Eq. (1.3) as a single complex field $(\phi_1, \phi_2) = \sqrt{2}\,(\mathrm{Re}\Psi, \mathrm{Im}\Psi)$. We introduce the gauge field A_μ through minimal coupling $\partial_\mu \Psi \to \left(\partial_\mu + ie^* A_\mu\right)\Psi$ for a field Ψ carrying charge e^*.

The current is obtained by differentiating the action with respect to A_μ, *viz.*

$$\begin{aligned} \langle J_\mu \rangle &= \frac{\delta S(A)}{\delta A_\mu} \\ &= ie^*\langle \Psi^* \partial_\mu \Psi - \Psi \partial_\mu \Psi^* \rangle + 2e^{*2} A_\mu \langle |\Psi|^2 \rangle , \end{aligned} \tag{2.4}$$

from which we derive the response function:

$$\begin{aligned} \Pi_{\mu\nu}(x, x') &= \frac{\delta}{\delta A_\nu(x')} \langle J_\mu(x) \rangle \Big|_{A=0} \\ &= \langle J_\mu(x)\, J_\nu(x') \rangle + 2e^{*2} \langle\, |\Psi|^2 \,\rangle\, \delta_{\mu\nu}\, \delta(x - x') . \end{aligned} \tag{2.5}$$

The first term is the paramagnetic response kernel $\Pi^{\mathrm{P}}_{\mu\nu}(x, x') = \langle J_\mu(x)\, J_\nu(x') \rangle$, and the second term is the diamagnetic response. The conductivity is then given by

$$\sigma(i\omega_m) = -\frac{1}{\omega_m}\, \Pi_{xx}(i\omega_m, q = 0) . \tag{2.6}$$

As in Eq. (2.2), the real frequency dynamics is obtained by analytic continuation,

$$\sigma(\omega) = \sigma(i\omega_m \to \omega + i\epsilon) . \tag{2.7}$$

Remarkably, in 2+1 dimensions the scaling dimension of the conductivity is zero [13]. As a result, near the critical point the conductivity has the scaling form [13, 14],

$$\sigma(\omega) = \sigma_q \, \Sigma_{\pm}(\omega/\Delta) \,. \tag{2.8}$$

Here $\sigma_q = e^{*2}/h$ is the quantum of conductance and $\Sigma_{\pm}$ are dimensionless universal functions of ω/Δ for the disordered (+), and ordered (−) phases.

2.3 Model and Methods

In order to simulate the continuum field theory Eq. (1.3) we consider the following discrete lattice model:

$$\begin{aligned} \mathcal{Z} &= \int \mathcal{D}\vec{\phi} e^{-S[\vec{\phi}]} \\ S &= \sum_{\langle ij \rangle} \vec{\phi}_i \cdot \vec{\phi}_j + \mu \sum_i |\vec{\phi}_i|^2 + g \sum_i |\vec{\phi}_i|^4 \,. \end{aligned} \tag{2.9}$$

Here $\vec{\phi}$ is an N component scalar field, residing on the sites of cubic lattice of linear size L with periodic boundary conditions. The model is the same as that considered in Ref. [15], as seen by rescaling $\vec{\phi}_i \to g^{-1/2}\vec{\phi}_i$. The long wavelength properties of Eq. (2.9) are captured by the field theory Eq. (1.3). This model can be interpreted either as a quantum mechanical partition function in discrete $2+1$ Euclidean space-time dimensions, or as a classical statistical mechanics model in three dimensions. Near the phase transition between ordered and disordered phases, this minimal model captures the critical properties of Eq. (1.3) while explicitly treating space and time on an equal footing and preserving exact particle-hole symmetry ($\Psi \to \Psi^*$) for the $N = 2$ case.

Next we define the discrete lattice version of the continuum observables. The scalar susceptibility is given by

$$\chi_s(\tau) = \sum_{x,y} \langle \vec{\phi}^2_{x,y,\tau} \, \vec{\phi}^2_{\mathbf{0}} \rangle - \langle \vec{\phi}^2_{\mathbf{0}} \rangle^2 \,. \tag{2.10}$$

To define the conductivity it is easier to consider the $U(1)$ symmetric complex field analog model of the $N = 2$ scalar field,

$$\begin{aligned} \mathcal{Z} &= \int \mathcal{D}\Psi \, \mathcal{D}\Psi^* \, e^{-S[\Psi,\Psi^*]} \\ S &= \sum_{\langle ij \rangle} \left(\Psi_i^* \Psi_j + \Psi_i \Psi_j^* \right) + 2\mu \sum_i |\Psi_i|^2 + 4g \sum_i |\Psi_i|^4 \,. \end{aligned} \tag{2.11}$$

We introduce the gauge field $A_\mu(i)$ through Peierls substitution $\Psi_i^* \Psi_{i+\mu} \rightarrow \Psi_i^* \Psi_{i+\mu} e^{ie^* A_\mu(i)}$. The current is then

$$J_\mu(i) = \frac{\delta S}{\delta A_\mu(i)} = ie^* \langle \Psi_i^* \Psi_{i+\hat{\mu}} e^{ie^* A_\mu(i)} - \text{c.c.} \rangle \tag{2.12}$$

and the response function,

$$\begin{aligned} \Pi_{\mu\nu}(i,j) &= \frac{\delta}{\delta A_\nu(j)} \langle J_\mu(i) \rangle \Big|_{A=0} \\ &= \Pi^{\mathrm{P}}_{\mu\nu}(i,j) + K \delta_{\mu\nu} \delta_{i,j} \ . \end{aligned} \tag{2.13}$$

$\Pi^{\mathrm{P}}_{\mu\nu}(i,j) = \langle J_\mu(i) \, J_\nu(j) \rangle$ and $K = -e^{*2} \langle \Psi_i^* \Psi_{i+\hat{\mu}} + \text{c.c.} \rangle$ are the lattice versions of, respectively, the paramagnetic and the diamagnetic response.

The simplicity of our model allowed us to simulate large system sizes, up to $L = 200$. Considering such large systems enabled us to accurately track the critical properties near the QCP. This is especially important in the ordered phase where the system is gapless and the dynamical response functions have power-law behavior. We implemented the highly efficient "worm algorithm" [16], sampling from a dual closed loops representation. The correlation time of the worm algorithm scales well with system size, suppressing the critical slowing down near the transition. We also extend the work of Ref. [16] to treat general O(N) models with $N > 2$. Details of the QMC algorithm can be found in appendix A. We compared our numerical results against previous QMC studies of O(N) models [17, 18] and with analytically solved limits and found good agreement within error bars.

A key ingredient of our analysis is the numerical analytic continuation of imaginary time QMC data to real frequency spectral functions. To do so we have to invert the relation,

$$\mathcal{G}(i\omega_m) = \int_0^\infty \frac{d\nu}{\pi} \frac{2\nu}{\omega_m^2 + \nu^2} A(\nu) \ . \tag{2.14}$$

Here $\mathcal{G}(i\omega_m)$ is a correlation function in Matsubara frequency space, evaluated by the QMC simulation, and $A(\nu)$ is the spectral function. However, the kernel has very small singular value eigenvalues, and the inversion can unwittingly amplify the statistical QMC noise in $\mathcal{G}(i\omega_m)$. A detailed discussion of methods which can circumvent these artifacts is presented in Appendix B.

2.4 Critical Energy Scales

2.4.1 Determination of the Critical Coupling

In order to study critical properties it is necessary to locate the QCP with high accuracy. We determine the critical coupling by finite size scaling analysis of the helicity modulus of the 2 + 1-dimensional quantum model. The helicity modulus Υ is defined by $\Upsilon \equiv \frac{1}{L}\frac{\partial^2 \ln \mathcal{Z}(\varphi)}{\partial \varphi^2}|_{\varphi=0}$ where $\mathcal{Z}(\varphi)$ is the partition function in the presence of a uniform phase twist φ. Near the critical point, ΥL is a universal constant, with only next-to-leading order corrections in the system size L [13, 19]. The critical coupling is then determined from the crossing point of $L\Upsilon$ for a sequence of increasing system sizes L. Illustrative examples for $N = 2$ and $N = 3$ are shown in Fig. 2.1. Curves for different system sizes cross at a single point with little variation with system size, allowing us to determine the critical coupling accurately.

We studied a few different parameter sets (g_c, μ_c) which are shown in Table 2.1. The use of multiple sets of model parameters for $N = 2$ allowed us to test the universality of our results. In most cases we tuned the transition by varying g, except in the case of dynamical conductivity, where we varied μ.

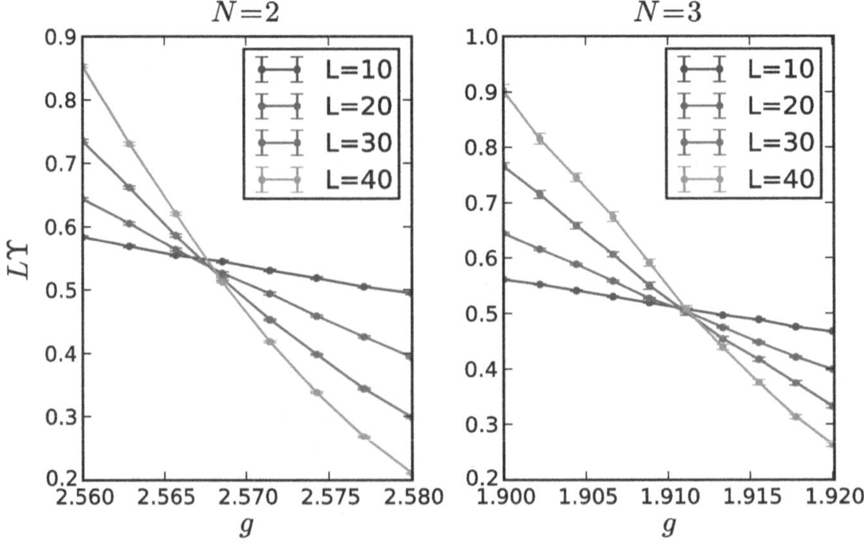

Fig. 2.1 *Curves* of $L\Upsilon$ for an sequence of increasing system size L for O($N = 2, 3$) models. The *curve* cross at a single point, from which we determine the value of g_c. Here we take $\mu = -0.5$ and $g_c = 2.568(2)$ for the $N = 2$ case and $g_c = 1.912(2)$ for $N = 3$

Table 2.1 List of model parameters studied, along with their critical couplings

Model	N	Model parameters	Critical coupling
A	2	$\mu = -0.5$	$g_c = 2.568(2)$
B	2	$\mu = -2$	$g_c = 3.908(2)$
C	2	$g = 7.6923$	$\mu_c = -5.883(2)$
D	3	$\mu = -0.5$	$g_c = 1.912(2)$
E	4	$\mu = -0.5$	$g_c = 1.516(2)$

2.4.2 Excitation Gap in the Disordered Phase

The gap in the disordered phase provides a reference energy scale for all dynamical properties. It can be extracted with high precision from the zero momentum two point Green's function [20],

$$G(\tau) = \sum_{x,y} \langle \vec{\phi}_{x,y,\tau} \cdot \vec{\phi}_{\mathbf{0}} \rangle , \tag{2.15}$$

without recourse to analytic continuation. At large imaginary times, $G(\tau)$ is expected to behave as

$$G(\tau) \sim e^{-\Delta\tau} + e^{-\Delta(\beta-\tau)} . \tag{2.16}$$

The gap Δ is evaluated by a fit to the above functional form. The evolution of the gap near the QCP is depicted in Fig. 2.2 for $N = 2, 3$. The gap softens as $\delta g \to 0$ according to the scaling form $\Delta(g) \sim \Delta_0 (\delta g)^\nu$, from which we extract Δ_0. For the correlation length exponent ν, we use values determined in previous high accuracy simulations [17, 18]: $\nu_2 = 0.6723(3)$, $\nu_3 = 0.710(2)$, and $\nu_4 = 0.749(2)$ for $N = 2$, $N = 3$, and $N = 4$ respectively.

We validated our results by performing a similar analysis of the long imaginary time form of the scalar susceptibility [15] $\chi_s(\tau) \sim \tau^{-1} e^{-2\tau\Delta}$. We found good agreement between the two approaches.

2.4.3 Helicity Modulus in the Ordered Phase

In two spatial dimensions, the helicity modulus is an energy scale that can be used to characterize the ordered phase. For $N = 2$ ($N = 3$) it plays the role of the superfluid stiffness (spin stiffness). Similarly to the gap in the disordered phase, the helicity modulus near the QCP vanishes according to the scaling behavior $\Upsilon = \Upsilon_0(\delta g)^\nu$. The ratio Υ_0/Δ_0 is universal. We find $\Upsilon_0/\Delta_0 = 0.44(1)$ for $N = 2$ and $\Upsilon_0/\Delta_0 = 0.34(1)$ for $N = 3$. This universal ratio was also calculated by means of non-pertubative renormalization group methods in Ref. [21].

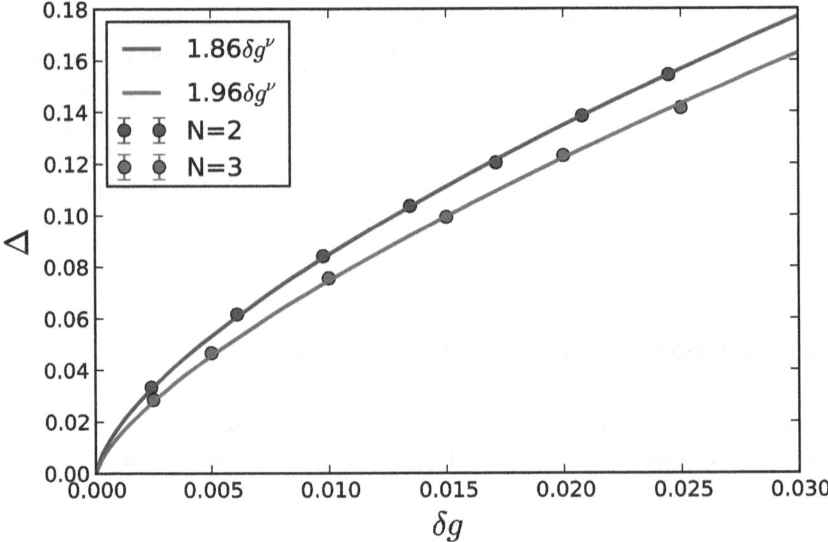

Fig. 2.2 Scaling of the gap $\Delta(\delta g)$ in the disordered phase for $N = 2, 3$ and $\mu = -0.5$. Fitting to the scaling form $\Delta = \Delta_0(\delta g)^\nu$ gives $\Delta_0 = 1.86(1)$ for $N = 2$, $\mu = -0.5$ and $\Delta_0 = 1.96(1)$ for $N = 3$, $\mu = -0.5$. Error bars are smaller than the symbols

2.5 Scalar Susceptibility

In the following, the universal scaling functions of the scalar susceptibility are computed for both phases (Fig. 2.3).

2.5.1 Matsubara Frequency Universal Scaling Function

In Fig. 2.4a numerical results for the $N = 2$ scalar susceptibility $\chi_s(i\omega_m)$ as a function of Matsubara frequency are presented for both phases. The scaling form Eq. (2.3) applies also to the correlation function in Matsubara space. The universal scaling function $\Phi(i\omega_m)$ is then computed by rescaling the $\chi_s(i\omega_m)$ curves according to Eq. (2.3). The collapse requires the extraction of the non-universal real constant C, which is expected to be a smooth function of δg. We find C by fitting $\chi_s(i\omega_m)$ at small ω_m to a polynomial in δg, and then subtracting it from $\chi_s(i\omega_m)$. The ω axis is then rescaled by Δ and the vertical axis is rescaled by $\Delta^{3-2/\nu}$.

Figure 2.4b, c shows the scaling procedure for $N = 2, 3$. The curves collapse into two universal functions $\Phi_\pm(i\omega_m)$. To test the universality of our results we repeated the scaling analysis at a different crossing point of the phase transition for the $N = 2$ case. The results are presented in Fig. 2.4b. The scaled curves for both sets of critical

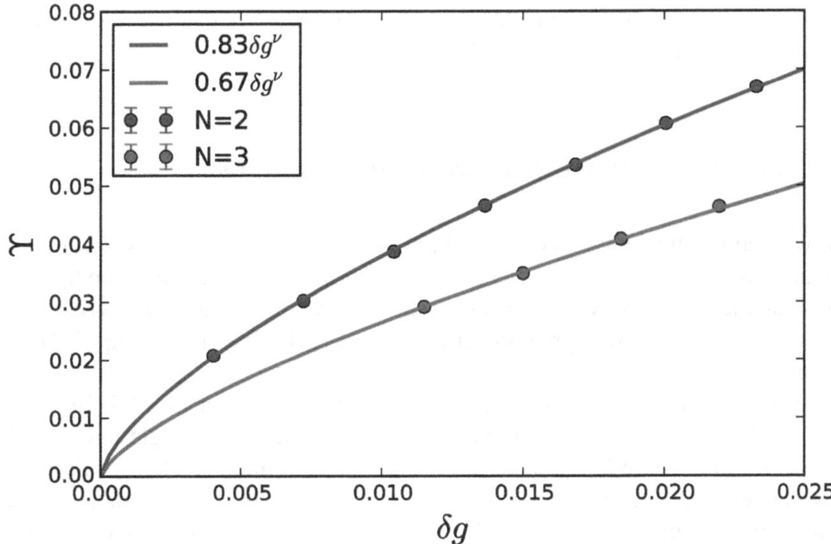

Fig. 2.3 Scaling of the helicity modulus $\Upsilon(\delta g)$ in the ordered phase for $N = 2, 3$ and $\mu = -0.5$. Fitting to the scaling form $\Upsilon = \Upsilon_0(\delta g)^{\nu}$ gives Υ_0=0.83(1) for $N = 2$, $\mu = -0.5$ and Υ_0=0.67(1) for $N = 3$, $\mu = -0.5$. Error bars are smaller than the symbols

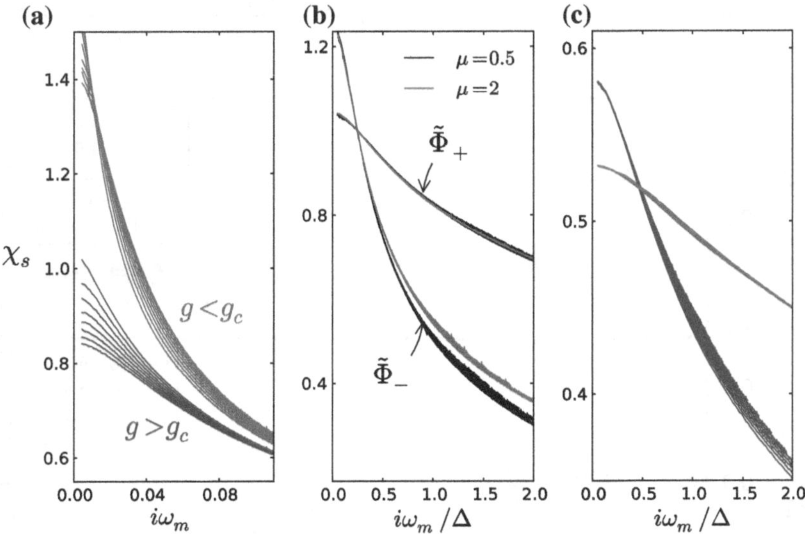

Fig. 2.4 **a** The scalar susceptibility $\chi_s(i\omega_m)$ for $N = 2$. The curves correspond to different values of δg below and above the phase transition. **b, c** universal scaling function after rescaling for $N = 2, 3$. In (**b**) we show the scaling function for two crossing points of the phase transition. The two rescaled *curves* agree very well, especially at low frequencies. Simulations were performed with $\mu = 0.5$ and $\mu = 2$ for $N = 2$ and $\mu = 0.5$ for $N = 3$

couplings agree very well, especially for low frequencies. This provides a stringent test for the consistency of our analysis.

2.5.2 Real Frequency Universal Scaling Function

Next we examine the imaginary part of the retarded response function $\chi_s''(\omega)$ obtained from analytic continuation of $\chi_s(i\omega_m)$. To extract the universal part of the line shape we rescale the ω axis by Δ and the vertical axis by $\Delta^{3-2/\nu}$. Note that this rescaling is done without any free fitting parameters, since the real constant C in Eq. (2.3) drops out from the spectral function.

The rescaled line shape in the ordered phase is shown in Fig. 2.5 for $N = 2$ and $N = 3$. Curves for different values of δg collapse into a single universal line shape especially at low frequencies. The line shape contains a clear peak that can be associated with the Higgs mode. Our analysis demonstrates that the Higgs peak is a universal feature in the spectral function that survives as a resonance arbitrarily close to the critical point.

Some universal values can be obtained by this analysis. For example, we consider the ratio between the Higgs mass in the ordered phase, defined by the maximum in $\chi_s''(\omega)$, and the gap in the disordered phase at mirror points across the transition. This ratio is found to be $m_H/\Delta = 2.1(3)$ and $m_H/\Delta = 2.2(3)$ for $N = 2$ and $N = 3$

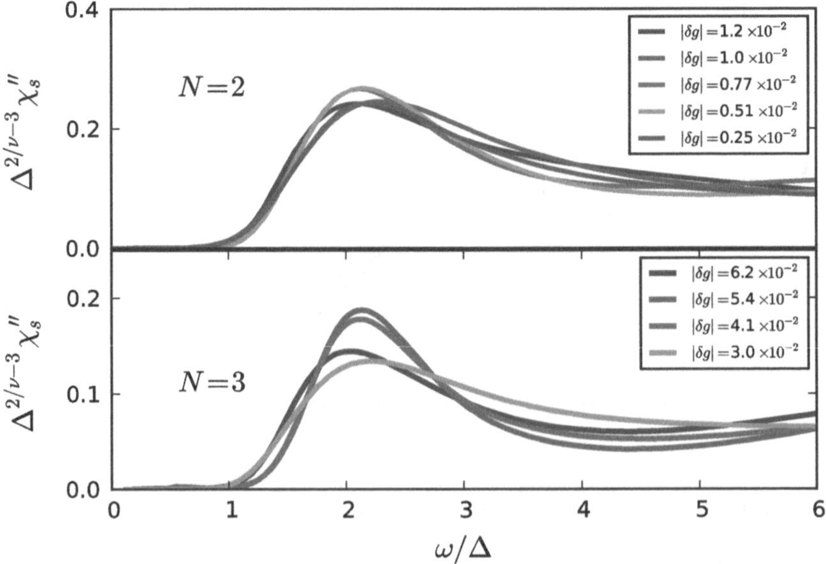

Fig. 2.5 $\chi_s''(\omega)$ in the ordered phase for $N = 2$ and 3. We scale the *curves* according to Eq. (2.3) for a range of tuning parameters δg near the critical point

respectively. We also obtain the fidelity $F = m_H/\Gamma$, where Γ is the full width at half-maximum. We measure Γ with respect to the leading edge at low frequency, since at low frequencies there is less contamination from the high frequency non-universal spectral weight. Since the entire functional form of the line shape is universal, F is a universal constant that characterizes the shape of the peak. We find $F = 2.4(10)$ for $N = 2$ and $F = 2.2(10)$ for $N = 3$.

The rescaled spectral function in Fig. 2.5 shows higher variability at high frequencies than at low frequencies. We attribute this to contamination from the non universal part of the spectrum and to systematic errors introduced by the maximum entropy ("MaxEnt") regularization of the analytic continuation, which is noisy in this regime.

In Fig. 2.6 we plot the rescaled line shape in the disordered phase for $N = 2$. The universal spectral function is gapped for $\omega < 2\Delta$ and rises sharply above the threshold. This behavior is in accordance with analytic predictions [12] and with previous QMC numerical simulation [22]. Previous studies found a Higgs-like resonance in the disordered phase above the threshold [22, 23]. However, we find that the peak seen in Fig. 2.6 at $\omega/\Delta \approx 3$ is very shallow relative to the background spectral weight. Thus we do not consider this to be conclusive evidence of a resonance. We note that numerical analytic continuation tends to produce oscillatory behavior near sharp features of the spectral function [24] and hence it is possible that the shallow peak might be an artifact of such an effect. For comparison, in Fig. 2.7 we show representative curves for the line shape on mirror points of the transition.

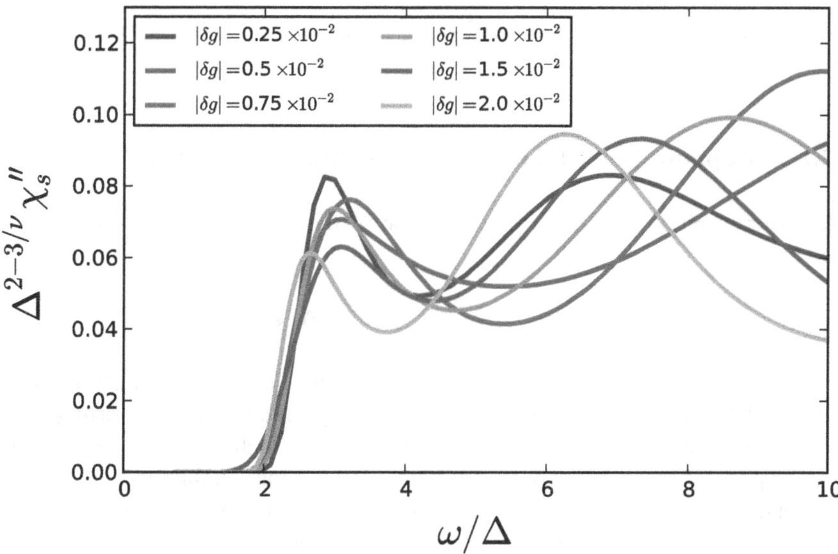

Fig. 2.6 $\chi_s''(\omega)$ in the disordered phase for $N = 2$. We scale the *curves* according to Eq. (2.3) for a range of tuning parameters δg near the critical point

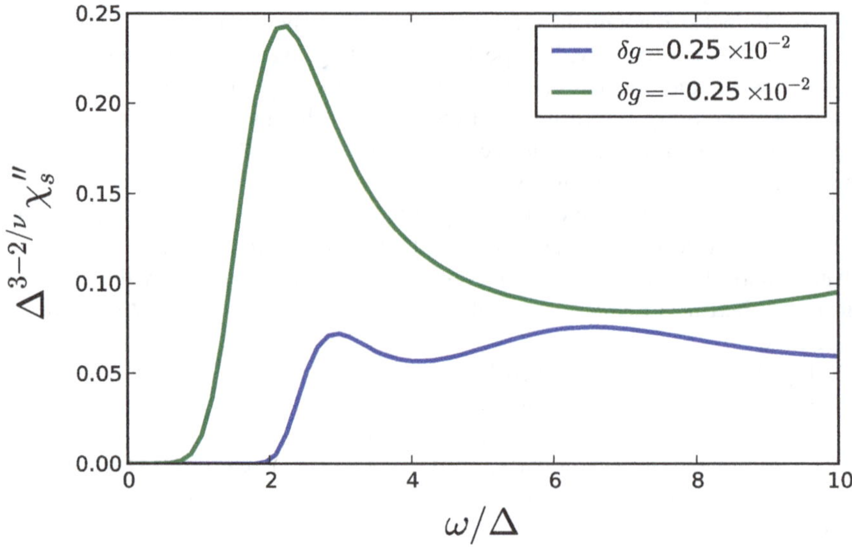

Fig. 2.7 Comparison of the scalar susceptibility line shape, $\chi_s''(\omega)$, on mirror points across the phase transition or $N = 2$. The *blue green curve* corresponds to disordered phase and the *green curve* to the ordered phase

If a resonance is at all present in the disordered phase, it is much less pronounced than in the ordered phase.

We summarize by plotting in Fig. 2.8 the critical energy scales near the QCP for $N = 2$ and 3.

2.5.2.1 Asymptotic Power Law Decay of the Scalar Susceptibility

In the ordered phase, the low frequency rise of the scalar susceptibility was predicted [1, 12, 25] to be

$$\Phi''_{-}(\omega) \sim (\omega/\Delta)^3 , \quad \omega \ll \Delta \ll 1. \tag{2.17}$$

The ω^3 rise is due to the decay of a Higgs mode into a pair of Goldstone modes. Equation (2.17) transforms into the large imaginary time asymptotic form $\chi_s(\tau) \sim 1/\tau^4$. Hence, to test Eq. (2.17) we examine the large τ behavior of $\chi_s(\tau)$. We note that this approach does not rely on analytic continuation, enabling us to study the low frequency dynamics in a numerically stable and well controlled manner.

In Fig. 2.9 we present $\chi_s(\tau)$ on a log-log plot for $N = 3, 4$ in the disordered phase with the detuning parameter $\delta g = 0.1 \times 10^{-2}$. For $N = 3, 4$ we indeed find agreement with the asymptotic behavior $\chi_s(\tau) \sim 1/\tau^4$ within the error bars. In Fig. 2.10 we present $\chi_s(\tau)$ for $N = 2$ on a log-log plot and on a semi-lrog plot.

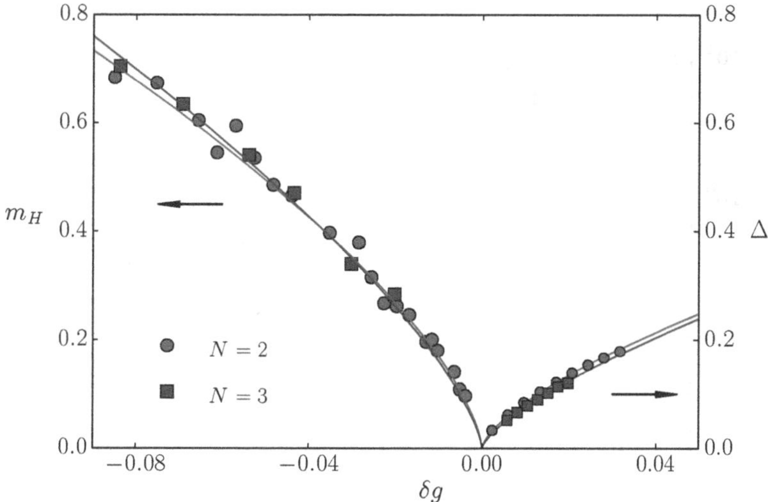

Fig. 2.8 Critical energy scales near the quantum phase transition in relativistic O(N) field theory for $N = 2, 3$. $\delta g \equiv (g - g_c)/g_c$ is the dimensionless tuning parameter. m_H is the Higgs peak energy (mass) in the ordered phase $\delta g < 0$, and Δ is the gap in the disordered phase $\delta g > 0$. Solid lines describe the critical behavior $m_H = \mathcal{B}_-|\delta g|^{\nu_N}$ and $\Delta = \mathcal{B}_+|\delta g|^{\nu_N}$

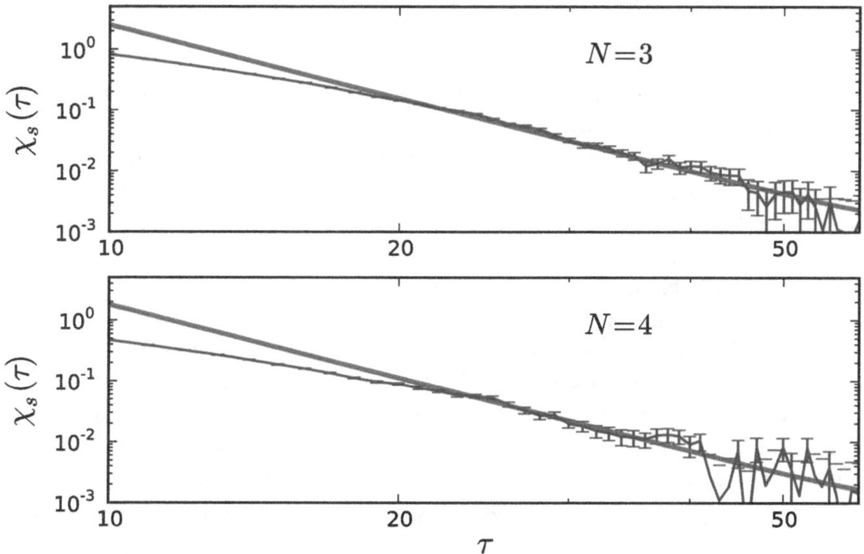

Fig. 2.9 Log-log scale plot for $\chi_s(\tau)$ in the ordered phase. For $N = 3, 4$ we indeed find the asymptotic behavior $\chi_s(\tau) \sim 1/\tau^4$ to agree within the error bars

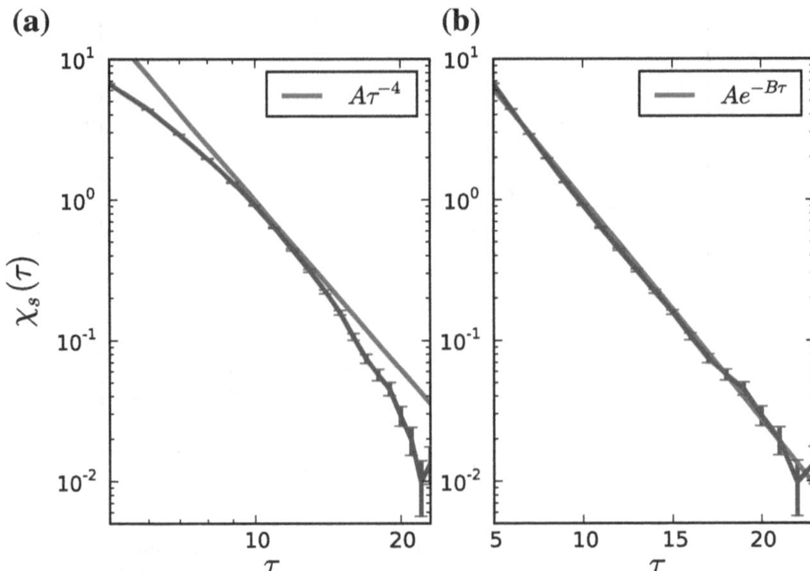

Fig. 2.10 $\chi_s(\tau)$ in the ordered phase for $N = 2$, plotted on a log-log scale in panel **a** and a semi-log scale in panel **b**. The *curve* deviates significantly from the expected $1/\tau^4$ power law form. Instead, the *curve* fits better to an exponential decay as in the disordered phase

Interestingly, for $N = 2$ we do not find a conclusive asymptotic fall-off as $1/\tau^4$. Instead, the data fits better to an exponential decay, as in the disordered phase. This indicates that the ω^3 sub-gap spectral weight, if at all present, is small compared to the spectral weight contained in the Higgs peak. Indeed we find excellent agreement between the large τ exponential decay rate and the value of m_H obtained from the MaxEnt analysis, further supporting our results for the Higgs mass. We note that a $1/\tau^4$ power law behavior might be regained at larger values of τ, but this lies below the statistical inference of our data.

Accurate determination of the scalar susceptibility at zero Matsubara frequency $\chi_s(i\omega = 0)$ is crucial for this analysis. Errors in $\chi_s(i\omega = 0)$ translate into an overall vertical shift of $\chi_s(\tau)$. This error can dominate the value of $\chi_s(\tau)$, especially at large τ where $\chi_s(\tau)$ is numerically small, and can lead to a bias in the power-law analysis. Typically, $\chi_s(i\omega = 0)$ is measured from a fluctuation relation $\chi_s(i\omega = 0) = \sum_{x,y,\tau} \langle \vec{\phi}^2_{x,y,\tau}\, \vec{\phi}^2_{\mathbf{0}} \rangle - \langle \vec{\phi}^2_{\mathbf{0}} \rangle^2$ and hence does not self-average [26] upon increasing the system size. To overcome this difficulty we computed $\chi_s(i\omega = 0)$ using a direct numerical derivative $\chi_s(i\omega = 0) = -d\langle \vec{\phi}^2 \rangle / d\mu$. To do so we evaluated $\langle \phi^2 \rangle$ for a set of values of μ within a narrow range $[\mu - \Delta\mu, \mu + \Delta\mu]$ and extracted the derivative by a polynomial fit in μ. We found that this method reduced the error in $\chi_s(i\omega = 0)$ by an order of magnitude and significantly improved the power law decay analysis.

2.6 Dynamical Conductivity

In Fig. 2.11 we present the dynamical conductivity in the disordered and ordered phases. In both cases the frequency axis ω is rescaled by Δ, noting that there is no need for a vertical rescaling since the conductivity is a universal amplitude. In both phases the curves collapse into single a universal shape, especially at low frequencies. The spectrum on the disordered side has a clear gap-like behavior up to a threshold frequency 2Δ. Beyond this threshold, the spectrum rises sharply and saturates at a universal value of $\sigma_{\mathrm{dis}}(\omega \gg \Delta) \approx 0.35(5)\,\sigma_{\mathrm{q}}$, where $\sigma_{\mathrm{q}} = e^{*2}/h$ is the quantum of conductance. These results should be compared with the line shape calculated diagrammatically in Ref. [14],

$$\sigma_{+}(\omega) = 2\pi\sigma_{\mathrm{q}} \left(\frac{\omega^2 - 4\Delta^2}{16\omega^2} \right) \Theta(\omega - 2\Delta)\,. \tag{2.18}$$

Similarly, in the ordered phase, the dynamical conductivity grows rapidly starting at a threshold frequency $\approx 2\Delta$, and saturates at high frequency at a value $\sigma_{\mathrm{ord}}(\omega \gg \Delta) \approx 0.25(5)\,\sigma_{\mathrm{q}}$. A calculation to leading order in weak coupling predicts [1, 27] (see also appendix C)

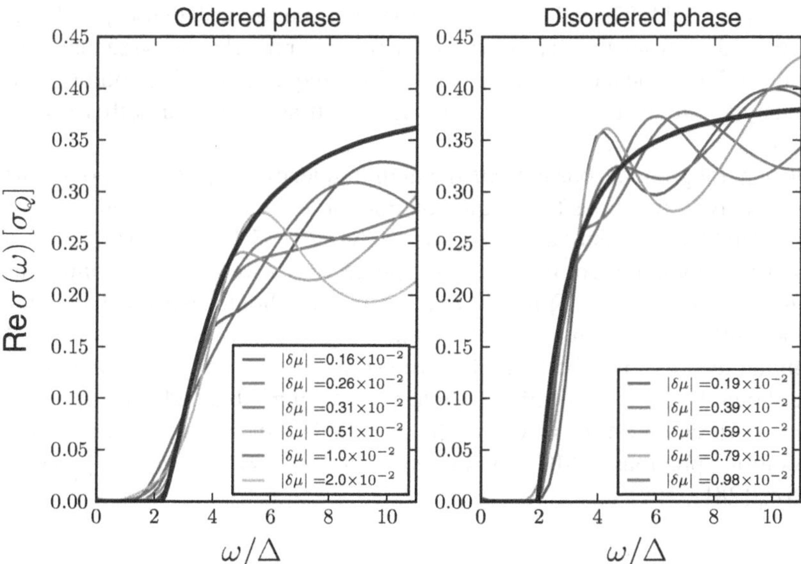

Fig. 2.11 The optical conductivity, $Re\sigma(\omega)$ in the ordered and disordered phases for $N = 2$. *Curves* are scaled by according to Eq. (2.8) for several values of the quantum tuning parameter δg near the critical point. The *solid black curves* show the analytic results from Refs. [1] and [14]

$$\sigma_{-}(\omega) = 2\pi\sigma_{\mathrm{q}} \left(\frac{\omega^2 - m_H^2}{4\omega^2} \right)^2 \Theta(\omega - m_H) . \tag{2.19}$$

In contrast to the disordered phase, there is a sub-gap component to the conductivity, owing to the gaplessness of the Goldstone mode(s). This feature is first evident at two loop order in a perturbative calculation of the conductivity. This was computed in Ref. [1], where it was found that the corresponding sub-threshold ($\omega < m_H$) contribution to $\sigma(\omega)$ is

$$\sigma_{-}(\omega)\Big|_{\omega<m_H} = \sigma_{\mathrm{q}} \cdot \frac{g m_H}{2^8 \pi} \left\{ \frac{N-2}{N} \left(\frac{16\omega}{15 m_H} + \frac{32\omega^3}{105 m_H^3} \right) + \right. \\ \left. + \frac{3N-5}{N} \frac{16\omega^5}{315 m_H^5} + \cdots \right\} + \mathcal{O}(g^2) . \tag{2.20}$$

Remarkably, for $N = 2$, the two leading order frequency terms in the sub-threshold conductivity vanish, resulting in a pronounced pseudogap behavior. Our numerical results appear to be qualitatively consistent with this analytic prediction. However, the coefficient of the leading ω^5 term is small, given by $3.2 \times 10^{-5}\, g/m_H^4$, and is not resolved within our numerical accuracy.

For comparison, the analytic curves corresponding to Eqs. (2.18) and (2.19) are plotted in Fig. 2.11. The value of m_H was taken from the scalar susceptibility analysis [15] and Δ from the gap analysis. There is a remarkable agreement between analytic and numerical curves especially at low frequencies. It is important to notice that analytic curves are presented without any fitting parameters (after setting m_H and Δ).

On general grounds, one expects the high frequency ($\omega \gg \Delta$) limit of the universal conductivity functions to be equal on both ordered and disordered phases. Here we find slightly different values, $\sigma_{\mathrm{dis}}(\omega \gg \Delta) \approx 0.35(5)\, \sigma_{\mathrm{q}}$ and $\sigma_{\mathrm{ord}}(\omega \gg \Delta) \approx 0.25(5)\, \sigma_{\mathrm{q}}$, although there is significant spread which we attribute to limitations of the analytic continuation. This high frequency value should also match the universal conductivity in the quantum critical regime at high frequencies ($\Delta = 0$ and $\omega \gg T$). Taking an average over both results, we estimate $\sigma_c^*(\omega \gg T) \approx 0.3(1)\, \sigma_{\mathrm{q}}$. This value should be compared with the value $\sigma_c^* = 0.39\, \sigma_{\mathrm{q}}$ obtained in the large N limit in Ref. [14], and with $\sigma_c^* = 0.251$ [28] obtained from leading correction in $1/N$. In addition, previous QMC simulations found $\sigma_c^* = 0.33\, \sigma_{\mathrm{q}}$ [19] and $\sigma_c^* = 0.285\, \sigma_{\mathrm{q}}$ [28].

2.6.1 Effect of Coulomb Interactions

Josephson junction arrays and granular superconducting films can often be described by charged lattice bosons [29], which interact at long range via e^2/r Coulomb interactions. When Coulomb interactions are present, the O(N) model Lagrangian should be augmented by a contribution

$$\Delta L = \int d^2x \, \text{i}n \frac{\partial \varphi}{\partial \tau} + \frac{1}{2} \int d^2x \int d^2x' n(\boldsymbol{x}) \frac{e^{*2}}{|\boldsymbol{x} - \boldsymbol{x}'|} n(\boldsymbol{x}') \tag{2.21}$$

where φ is the phase of the order parameter. We parameterize the $\vec{\phi}$ field in terms of longitudinal (σ) and transverse (π) fluctuations:

$$\vec{\phi} = (\phi_0 + \sigma \, , \, \pi) \, , \tag{2.22}$$

where $\phi_0 \equiv \left|\langle \vec{\phi} \rangle\right|$. To lowest order, we have $\varphi = \pi/\eta\sqrt{N}$, where $\eta \equiv \phi_0/\sqrt{N}$ is proportional to the magnitude of the order parameter. Integrating out the density field $n(\boldsymbol{x}, \tau)$, we find that the π propagator becomes

$$G_{\pi\pi}(q) = \frac{1}{q_0^2 + \boldsymbol{q}^2 + \alpha \, |\boldsymbol{q}| \, q_0^2} \, , \tag{2.23}$$

where $\alpha = \eta g \hbar v/\pi e^{*2}$ and v is the velocity ('speed of light') in the original O(N) model. This new π-field propagator has a 2D plasmon pole located at $q_0 = \sqrt{-|\boldsymbol{q}|/\alpha}$ for small $\boldsymbol{q}$. Plugging this into the expression for the electromagnetic kernel, in Eq. (E1) of Ref. [1], we find, to order g^0,

$$\sigma(\omega) = 2\sigma_{\text{q}} \left(\frac{\alpha}{m_H}\right)^2 (\omega - m_H)^4 \; > \Theta(\omega - m_H) \, . \tag{2.24}$$

Thus, the dynamical conductivity of two dimensional superconductors rises above the Higgs threshold with a modified power law $\sigma(\omega) \propto (\omega - m_H)^4$.

2.7 Discussion and Summary

In this chapter we studied the critical dynamical properties of O(N)-symmetric models with relativistic dynamics in two space dimensions. In particular we computed the line shape of the scalar susceptibility and the optical conductivity on either side of the quantum phase transition. Our results focus on properties that are universal in nature and are therefore relevant for many experimental realizations of quantum phase transitions.

We showed that the scalar susceptibility, in the ordered phase, contains a clear resonance at the Higgs mass m_H. By contrast, in the disordered phase the scalar susceptibility has a threshold at $\omega = 2\Delta$ with no conclusive evidence for a resonance above the threshold. In addition we provide two universal dimensionless constants that characterize the dynamics: the ratio between the Higgs mass and the single particle gap on mirror points across the transition, and the fidelity of the Higgs resonance. These predictions could be tested by future, high resolution, experiments of the superfluid to Mott insulator transition in cold atomic lattices [2].

It is important to note that, close to the critical point, the scalar susceptibility captures the low frequency behavior of a generic experimental probe that couples to the order parameter amplitude and not to its direction [15].

We have also presented results for the optical conductivity on both sides of the phase transition. In both cases we find a sharp rise of the spectral function at $\omega \approx 2\Delta$. The threshold frequency in the ordered phase can be associated with the Higgs mass m_H. This provides an independent estimate of the Higgs mass, one which agrees very well with the value obtained from the scalar susceptibility analysis. In addition we have computed the high frequency ($\omega \gg T$) universal conductance $\sigma_c^* = 0.3(\pm 0.1) \times \sigma_q$. This value is with agreement with previous analytic calculations [14]. Unfortunately the low frequency ("hydrodynamic") limit $\omega \ll T$ is not accessible in the QMC simulation, as was discussed in Ref. [14].

Finally we have shown that for charged system with Coulomb interaction the power law of the spectral rise above the threshold changes from 2 to 4.

We hope that our results will motivate measurements of the optical conductivity in cold atoms by optical lattice phase modulation, as was suggested in Ref. [10]. Such experiments could accurately measure the universal optical conductivity near the QCP and even the universal resistivity right at the critical point. Our analysis may also shed light on recent experiments on the superconductor to insulator transition in granular superconductors [11]. In this context will be interesting to extend these calculations to systems with varying degrees of disorder.

References

1. D. Podolsky, A. Auerbach, D.P. Arovas, Visibility of the amplitude (Higgs) mode in condensed matter. Phys. Rev. B **84**(17), 174522 (2011). doi:10.1103/PhysRevB.84.174522. http://link.aps.org/doi/10.1103/PhysRevB.84.174522
2. M. Endres et al., The 'Higgs' amplitude mode at the two-dimensional super fluid/Mott insulator transition. Nature **487**(7408), 454–458 (2012). ISSN 0028-0836. doi:10.1038/nature11255. http://www.nature.com/nature/journal/v487/n7408/full/nature11255.html Accessed 11 Oct 2012
3. J. Demsar, K. Biljakovi, D. Mihailovic, Single particle and collective excitations in the one-dimensional charge density wave solid K0:3MoO3 probed in real time by femtosecond spectroscopy. Phys. Rev. Lett. **83**(4), 800–803 (1999). doi:10.1103/PhysRevLett.83.800. http://link.aps.org/doi/10.1103/PhysRevLett.83.800
4. R. Yusupov et al., Coherent dynamics of macroscopic electronic order through a symmetry breaking transition. Nat. Phys. **6**(9), 681–684 (2010). ISSN 1745-2473. doi:10.1038/nphys1738
5. K.B. Lyons et al., Dynamics of spin fluctuations in lanthanum cuprate. Phys. Rev. B **37**(4), 2353–2356 (1988). doi:10.1103/PhysRevB.37.2353. http://link.aps.org/doi/10.1103/PhysRevB.37.2353 (Bibliography 76)
6. J.B. Parkinson, Interaction of light with defects in antiferromagnetic perovskites. J. Phys. C: Solid State Phys. **2**(11), 2003 (1969). http://stacks.iop.org/0022-3719/2/i=11/a=315

7. P.A. Fleury, H.J. Guggenheim, Magnon-Pair modes in two dimensions. Phys. Rev. Lett. **24**(24), 1346–1349 (1970). doi:10.1103/PhysRevLett.24.1346. http://link.aps.org/doi/10.1103/PhysRevLett.24.1346
8. R.J. Elliott, M.F. Thorpe, The effects of magnon-magnon interaction on the two-magnon spectra of antiferromagnets. J. Phys. C: Solid State Phys. **2**(9), 1630 (1969). http://stacks.iop.org/0022-3719/2/i=9/a=312
9. B. Sriram Shastry, B.I. Shraiman, Theory of Raman scattering in Mott-Hubbard systems. Phys. Rev. Lett. **65**(8), 1068–1071 (1990). doi:10.1103/PhysRevLett.65.1068. http://link.aps.org/doi/10.1103/PhysRevLett.65.1068
10. A. Tokuno, T. Giamarchi, Spectroscopy for cold atom gases in periodically phase-modulated optical lattices. Phys. Rev. Lett. **106**(20), 205301 (2011). doi:10.1103/PhysRevLett.106.205301. http://link.aps.org/doi/10.1103/PhysRevLett.106.205301
11. D. Sherman et al., Effect of Coulomb interactions on the disorder-driven superconductor-insulator transition. Phys. Rev. B **89**(3), 035149 (2014). doi:10.1103/PhysRevB.89.035149. http://link.aps.org/doi/10.1103/PhysRevB.89.035149
12. D. Podolsky, S. Sachdev, Spectral functions of the Higgs mode near two-dimensional quantum critical points. Phys. Rev. B **86**(5), 054508 (2012). doi:10.1103/PhysRevB.86.054508. http://link.aps.org/doi/10.1103/PhysRevB.86.054508
13. M.P.A. Fisher, G. Grinstein, S.M. Girvin, Presence of quantum diffusion in two dimensions: universal resistance at the superconductor-insulator transition. Phys. Rev. Lett. **64**(5), 587–590 (1990). doi:10.1103/PhysRevLett.64.587. http://link.aps.org/doi/10.1103/PhysRevLett.64.587
14. K. Damle, S. Sachdev, Nonzero-temperature transport near quantum critical points. Phys. Rev. B **56**(14), 8714–8733 (1997). doi:10.1103/PhysRevB.56.8714. http://link.aps.org/doi/10.1103/PhysRevB.56.8714
15. S. Gazit, D. Podolsky, A. Auerbach, Fate of the Higgs mode near quantum criticality. Phys. Rev. Lett. **110**(14), 140401 (2013). doi:10.1103/PhysRevLett.110.140401. http://link.aps.org/doi/10.1103/PhysRevLett.110.140401
16. N. Prokof'ev, B. Svistunov, Worm algorithms for classical statistical models. Phys. Rev. Lett. **87**(16), 160601 (2001). doi:10.1103/PhysRevLett.87.160601. http://link.aps.org/doi/10.1103/PhysRevLett.87.160601 Accessed 26 Mar 2012. (Bibliography 77)
17. M. Hasenbusch, Eliminating leading corrections to scaling in the three-dimensional O (N)-symmetric 4 model: N = 3 and 4. J. Phys. A: Math. Gen. **34**(40), 8221–8236 (2001). issn: 0305–4470, 1361–6447, doi:10.1088/0305-4470/34/40/302. http://iopscience.iop.org/0305-4470/34/40/302 Accessed 18 June 2012
18. M. Hasenbusch, T. Torok, High-precision Monte Carlo study of the 3D XY -universality class. J. Phys. A: Math. Gen. **32**(36), 6361–6371 (1999). ISSN 0305-4470, 1361–6447, doi:10.1088/0305-4470/32/36/301. http://iopscience.iop.org/0305-4470/32/36/301 Accessed 18 June 2012
19. J. Smakov, E. Sorensen, Universal scaling of the conductivity at the super fluid-insulator phase transition. Phys. Rev. Lett. **95**(18), 180603 (2005). doi:10.1103/PhysRevLett.95.180603. http://link.aps.org/doi/10.1103/PhysRevLett.95.180603
20. B. Capogrosso-Sansone et al., Monte Carlo study of the two-dimensional Bose-Hubbard model. Phys. Rev. A **77**(1), 015602 (2008). doi:10.1103/PhysRevA.77.015602. http://link.aps.org/doi/10.1103/PhysRevA.77.015602
21. A. Ran çon, N. Dupuis, Higgs amplitude mode in the vicinity of a (2 + 1)- dimensional quantum critical point. Phys. Rev. B **89**(18), 180501 (2014). doi:10.1103/PhysRevB.89.180501. http://link.aps.org/doi/10.1103/PhysRevB.89.180501
22. K. Chen et al., Universal properties of the higgs resonance in (2 + 1)-dimensional U(1) critical systems. Phys. Rev. Lett. **110**(17), 170403 (2013). doi:10.1103/PhysRevLett.110.170403. http://link.aps.org/doi/10.1103/PhysRevLett.110.170403
23. L. Pollet, N. Prokof'ev, Higgs mode in a two-dimensional super fluid. Phys. Rev. Lett. **109**(1), 010401 (2012). doi:10.1103/PhysRevLett.109.010401. http://link.aps.org/doi/10.1103/PhysRevLett.109.010401

24. K.S.D. Beach, Identifying the maximum entropy method as a special limit of stochastic analytic continuation. arXiv preprint (2004). cond-mat/0403055
25. S. Sachdev, Universal relaxational dynamics near two-dimensional quantum critical points. Phys. Rev. B **59**(21), 14054–14073 (1999). doi:10.1103/PhysRevB.59.14054. http://link.aps.org/doi/10.1103/PhysRevB.59.14054 Accessed 15 Oct 2012
26. A. Milchev, K. Binder, D.W. Heermann, Fluctuations and lack of self averaging in the kinetics of domain growth. English. Zeitschrift fr Phys. B Condens. Matter **63**(4), 521–535 (1986). issn: 0722–3277. doi:10.1007/BF01726202. (Bibliography 78)
27. N.H. Lindner, A. Auerbach, Conductivity of hard core bosons: a paradigm of a bad metal. Phys. Rev. B **81**(5), 054512 (2010). doi:10.1103/PhysRevB.81.054512. http://link.aps.org/doi/10.1103/PhysRevB.81.054512
28. M.-C. Cha et al, Universal conductivity of two-dimensional films at the superconductorinsulator transition. Phys. Rev. B **44**(13), 6883–6902 (1991). doi:10.1103/PhysRevB.44.6883. http://link.aps.org/doi/10.1103/PhysRevB.44.6883
29. A. Mihlin, A. Auerbach, Temperature dependence of the order parameter of cuprate superconductors. Phys. Rev. B **80**(13), 134521 (2009). doi:10.1103/PhysRevB.80.134521. http://link.aps.org/doi/10.1103/PhysRevB.80.134521

Chapter 3
Critical Conductivity and Charge Vortex Duality Near Quantum Criticality

3.1 Introduction

Three decades ago, Fisher and Lee [1] showed that the SIT can be described as a *Bose condensation* of quantum vortices. Despite the appeal of this description, ρ_v, the *vortex condensate stiffness*, has remained an elusive observable, for which no experimental probe has yet been proposed. Also, to our knowledge, ρ_v has not been calculated near the critical point, for any microscopic model.

In this chapter, we address this problem, by using the exact reciprocity relation, presented in Eq. (1.18), between complex *dynamical* conductivities of bosons (σ) and vortices (σ_v) at *finite* frequency:

$$\sigma(\omega) \times \sigma_v(\omega) = q^2/h^2 , \tag{3.1}$$

where q is the boson charge (=$2e$ in superconductors) [2]. At low frequencies, this equation is dominated by the *reactive* (imaginary) conductivities. The superfluid stiffness ρ_s in the superfluid phase can be measured by the low frequency inductance L_{sf}, $\rho_s = \hbar/(2\pi\sigma_q L_{sf})$, where $\sigma_q = q^2/h$ is the quantum of conductance. Equation (3.1) allows us to identify the elusive vortex condensate stiffness with the *capacitance per square* in the insulating phase, $C_{ins} = \hbar\sigma_q/(2\pi\rho_v)$.

We shall examine quantitatively the charge-vortex duality (CVD) hypothesis [2, 3], which relates the superfluid to the insulating phase. The strict CVD hypothesis predicts a universal critical conductivity at the SIT given by σ_q. Experiments, however, have measured non-universal values of the critical conductivity [4].

Several factors can spoil CVD in experiments (i) Bosons (in the superfluid) and vortices (in the insulator) have different masses and interaction ranges [5, 6]. (ii) Potential energy (both confining and disordered) couples differently to charges and vortices. (iii) In superconductors, fermionic (Bogoliubov) quasiparticles produce dissipation, which can alter the phase diagram from the purely bosonic theory.

For strict CVD to hold, one expects $\rho_s(-\delta g) = \rho_v(\delta g)$, where $\pm\delta g$ are mirror points on either side of the SIT. Figure 3.1 depicts all the critical energy scales of

S. Gazit, *Dynamics Near Quantum Criticality in Two Space Dimensions*, Springer Theses, DOI 10.1007/978-3-319-19354-0_3

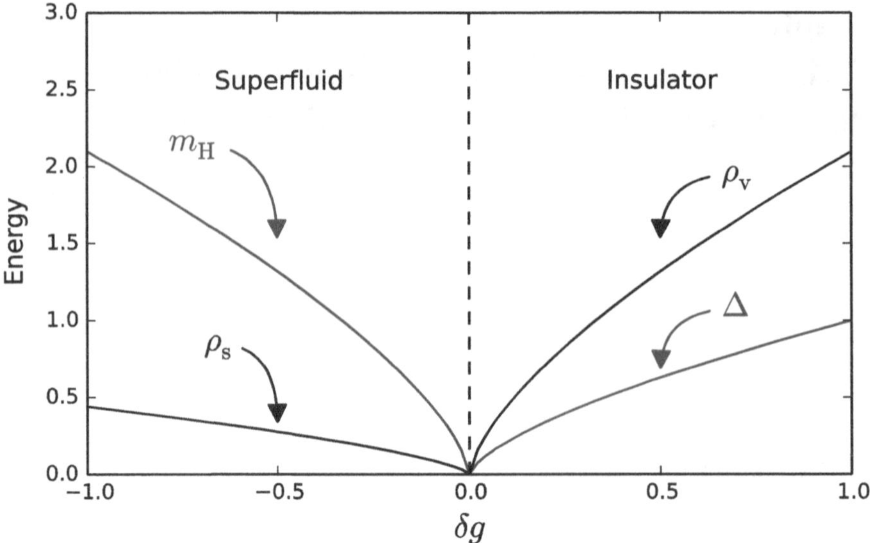

Fig. 3.1 Critical energy scales near the SIT computed by QMC. The superfluid is characterized by the mass of the amplitude mode, m_H, and the superfluid stiffness, ρ_{s}; the insulator by the single particle gap, Δ, and the vortex condensate stiffness, ρ_{v}. The amplitude ratios $m_H(-\delta g)/\Delta(\delta g) = 2.1(3)$, [7] $\rho_{\mathrm{s}}(-\delta g)/\Delta(\delta g) = 0.44(1)$ [8], and $\rho_{\mathrm{v}}(-\delta g)/\Delta(\delta g) = 2.1(1)$ are universal

the relativistic O(2) field theory, obtained by large-scale Monte Carlo simulation. In addition to the Higgs mass m_{H} and the charge gap Δ, which vanish at the critical point, we compare the energy scales ρ_{s} and ρ_{v} which are also critical, but have different relative amplitudes. The ratio $\rho_{\mathrm{s}}(-\delta g)/\rho_{\mathrm{v}}(\delta g) = 0.21(1)$. The deviation from unity quantifies the violation of CVD.

It is interesting to ask whether CVD is better satisfied at finite frequencies. To address this, we propose the product function

$$\mathcal{R}(z) \equiv \sigma(z, -\delta g) \times \sigma(z, \delta g)/\sigma_q^2, \tag{3.2}$$

as a measure of CVD between mirror points. Here, z denotes either a real or a Matsubara frequency. The high frequency conductivity (after removal of cut-off dependent corrections) reaches a universal value $\sigma^* = 0.355(5)\sigma_q$ [9, 10]. We compute the function $\mathcal{R}(i\omega_m)$ and address its implications to CVD. To supplement the numerical calculation, we compute the complex optical conductivity at one loop order in weak coupling. We conclude by proposing an experimental measure of the vortex condensate stiffness ρ_{v} for neutral bosons in an optical lattice.

3.2 Vortex Transport Theory

Boson charge current $\vec{j}$ is driven by an electrochemical field $\vec{E}$. Vortices are point particles in two dimensions. The vorticity current $\vec{j}_{\rm v}(t)$ is driven by the Magnus field $\vec{E}_{\rm v}$. Hydrodynamics dictate simple relations between electrochemical field and vortex number current, and between boson charge current and Magnus field [11]:

$$E_v^\alpha = \frac{h}{q}\epsilon^{\alpha\beta} j^\beta , \quad E^\alpha = \frac{h}{q}\epsilon^{\alpha\beta} j_v^\beta , \tag{3.3}$$

where $\epsilon = i\sigma^y$ is the two dimensional antisymmetric tensor. We note that Eq. (3.3) are *instantaneous*. Conductivity relates currents to their driving fields,

$$j_{(\rm v)}^\alpha(t) = \int_{-\infty}^t dt' \sigma_{(\rm v)}^{\alpha\beta}(t-t') E_{(\rm v)}^\beta(t') . \tag{3.4}$$

By Fourier transformation, the complex dynamical conductivities obey a reciprocity relation $\varepsilon^\top \sigma_{\rm v} \varepsilon = (q^2/h^2)\sigma^{-1}$. For the case of an isotropic longitudinal conductivity $\sigma^{xx} = \sigma^{yy} = \sigma$, one obtains the reciprocity Eq. (3.1), which can be analytically continued to Matsubara space $\omega \to i\omega_n$.

3.3 Model and Observables

For numerical simulations we study the discretized partition function presented in Eq. (2.9). $\mathcal{Z} = \int \mathcal{D}\vec{\phi}\mathcal{D}\vec{\phi}^* e^{-\mathcal{S}[\vec{\phi},\vec{\phi}^*]}$, where the real action S on Euclidean space-time is

$$\mathcal{S} = \sum_{\langle i,j\rangle} \vec{\phi}_i \vec{\phi}_j^* + \text{c.c} + 2\mu \sum_i |\vec{\phi}_i|^2 + 4g \sum_i |\vec{\phi}_i|^4 . \tag{3.5}$$

Here $\vec{\phi}_i$ are complex variables defined on a cubic lattice of size $L \times L \times \beta$. We take $\beta = L$ throughout. For $\mu < 0$, this model undergoes a continuous zero temperature quantum phase transition (QCP) between a superfluid with $\langle\vec{\phi}\rangle \neq 0$ for $g < g_c$ and an insulator with $\langle\vec{\phi}\rangle = 0$ for $g > g_c$. We define the quantum detuning parameter $\delta g = (g - g_c)/g_c$.

The critical energy scales near the SIT, as shown in Fig. 3.1, in the superfulid phase are the amplitude mode mass m_H and the superfluid stiffness $\rho_{\rm s}$ [7, 12]. In the insulating phase excitations are gapped, with single-particle gap Δ.

In the superfluid phase, the reactive conductivity diverges as $\operatorname{Im}\sigma_{\rm sf}(\omega) = 2\pi\sigma_q\rho_{\rm s}(-\delta g)/(\hbar\omega)$. Previous calculations [13] show that the dissipative component has a small sub-gap contribution below the Higgs mass, $0 < \omega \ll m_H(-\delta g)$ which goes as $\operatorname{Re}(\sigma_{\rm sf}(\omega)) \sim \omega^5$. This is negligible as $\omega \to 0$ and the analytic continuation to Matsubara frequency yields

$$\tilde{\sigma}_{\rm sf}(\omega_m) \sim \frac{2\pi\sigma_q\,\rho_{\rm s}}{\hbar\omega_m}, \quad (\text{for } \omega_m \ll m_H). \tag{3.6}$$

In the insulator, the dissipative conductivity vanishes identically below the charge gap $\Delta(\delta g)$ [8, 14]. The reactive conductivity vanishes linearly with frequency $\mathrm{Im}\,\sigma_{\rm ins}(\omega) = -C_{\rm ins}\omega$, where $C_{\rm ins}$ is the capacitance per square. Therefore, $\sigma_{\rm ins}$ can be analytically continued to Matsubara space as

$$\tilde{\sigma}_{\rm ins}(\omega_m) \sim C_{\rm ins}\,\omega_m\,, \quad (\text{for } \omega_m \ll \Delta). \tag{3.7}$$

Equation (2.8) implies that the capacitance $C_{\rm ins}$ diverges near the QCP as $C_{\rm ins} \sim 1/\Delta$. The capacitance measures the dielectric response of the insulator. Its divergence reflects the large particle-hole fluctuations near the transition.

In the vortex description the insulator is a bose condensate of vortices, with a low frequency vortex conductivity $\tilde{\sigma}_{\rm v}(\omega_m) = \rho_{\rm v}/(\hbar^2\omega_m)$. As a consequence, $\rho_{\rm v}$ can be defined in terms of the capacitance by applying Eq. (3.1),

$$\rho_{\rm v} \equiv \frac{\hbar\sigma_q}{2\pi C_{\rm ins}}. \tag{3.8}$$

We shall use this important relation to test for CVD in the 2+1 dimensional O(2) field theory.

3.4 Methods

A large scale QMC simulation of Eq. (3.5) is used to evaluate Eq. (2.6). We consider large systems, of linear size up to $L = 512$, which is crucial for obtaining universal properties. To validate the universality of our results we performed our analysis on two distinct crossing points of the SIT, by choosing $\mu = -0.5$ and $\mu = -5.89391$ and tuning g across the transition. We found excellent agreement within the error bars. Henceforth we will only present results for $\mu = -5.89391$, a value which has been argued to reduce finite size corrections to scaling [15].

3.5 Results

We locate the critical coupling $g_c(\mu)$, with high accuracy, by a finite size scaling analysis of the superfluid stiffness following the numerical method presented in Sect. 2.4.1. In this work we find $g_c = 7.0284(3)$. We extract the gap Δ in the insulator by analyzing the asymptotic large imaginary time decay of the two point Green's function as was explained in Sect. 2.4.2.

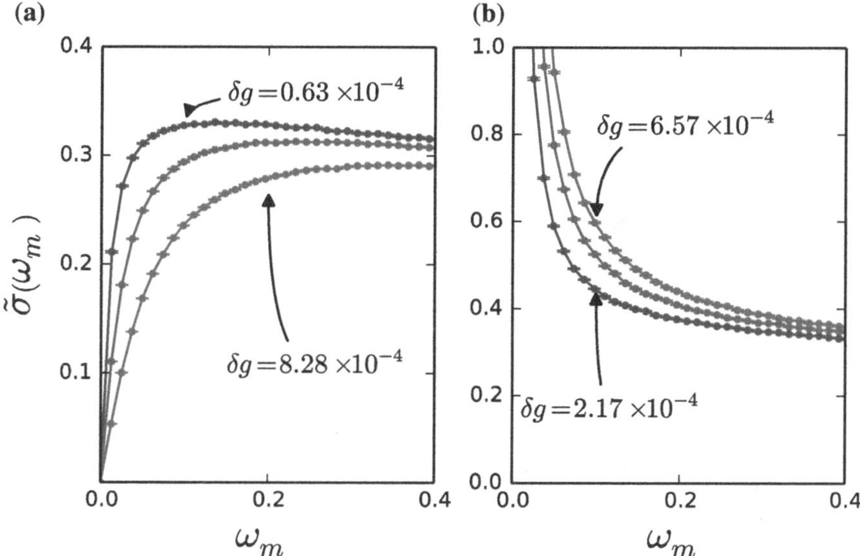

Fig. 3.2 The conductivity as a function Matsubara frequency. The *curves* differ by the detuning parameter δg. In the insulator, the low frequency conductivity is linear, $\sigma_{\text{ins}} \sim \omega_m$ indicating capacitive behavior. In the superfluid, the conductivity diverges as $\sigma_{\text{sf}} \sim 1/\omega_m$ indicating inductive response

In Fig. 3.2 we present the dynamical conductivity $\sigma(\omega_m)$ as a function of Matsubara frequency, both in the insulator and in the superfluid, for a range of detuning parameters δg near the critical point. To suppress finite size effects in the insulator we used an improved estimator, in which we consider only loop configurations with zero winding number [9, 10]. We find that the dynamical conductivity as a function of ω_m in Fig. 3.2, follows the form of the low frequency reactive conductivity both in the superfluid, Eq. (3.6), and in the insulator, Eq. (3.7).

Next we calculate ρ_{s} and ρ_{v} in their respective phases. The superfluid stiffness ρ_{s} was calculated using the standard method of winding number fluctuations [16]. In order to extract ρ_{v} we use the relation in Eq. (3.8). As a concrete Monte Carlo observable for the capacitance we use the conductivity evaluated at the first non-zero Matsubara frequency:

$$C(\delta g) = \lim_{L\to\infty} \frac{\sigma(\omega_m = \frac{2\pi}{L}, \delta g)}{\frac{2\pi}{L}}. \tag{3.9}$$

Both the vortex condensate stiffness ρ_{v} and the superfluid stiffness ρ_{s} near the critical point follow a power law behavior $\rho_{\{\text{s},v\}} \sim \rho^0_{\{\text{s},v\}} |\delta g|^{\nu}$. The non-universal prefactors ρ^0_{v} and ρ^0_{s} are extracted by a numerical fit.

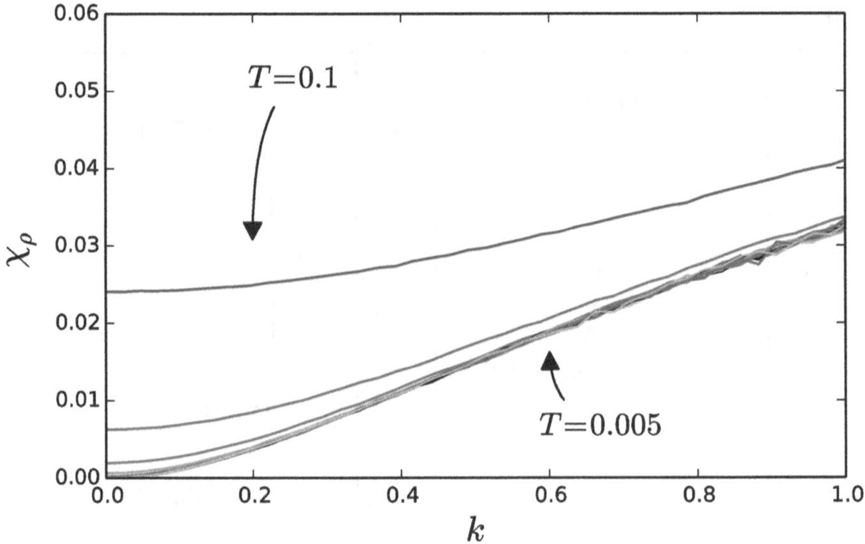

Fig. 3.3 The charge susceptibility $\chi_\rho(k, T)$ as a function of wave number k. We fix $\delta g = 1.12 \times 10^{-2}$, for which $\Delta = 0.103$. Each *curve* corresponds to a different temperature T. The temperature T and Δ are measured in units of the inverse lattice constant.

3.6 Charge Susceptibility and Finite Temperature Effects

The capacitance is a thermodynamic response as such it can measured by a static observable. To see that, we note that in the insulator, the current and charge response functions are related by the continuity equation, $\Pi_{xx}(k, \omega) = -\frac{\omega^2}{k^2}\chi_\rho(k, \omega)$, where $\chi_\rho(k, \omega)$ is the charge susceptibility. Hence,

$$C_{\text{ins}} = \lim_{\omega \to 0} \lim_{k \to 0} \frac{\Pi_{xx}(k, \omega)}{-\omega^2} = \lim_{k \to 0} \frac{\chi_\rho(k, \omega = 0)}{k^2}, \tag{3.10}$$

where the $\omega \to 0$ and $k \to 0$ limits commute since the insulator is gapped [17] and the charge operator in the O(2) model is defined as $\rho = i\,(\varphi^* \partial_t \varphi - \varphi \partial_t \varphi^*)$. Thus, the capacitance is simply related to the finite k compressibility of the Mott insulator.

Real experiments are performed at finite temperature and hence it is important to study the capacitance and the charge susceptibility $\chi_\rho(k, T)$ as a function of temperature.

As an illustrative example, in Fig. 3.3 we depict $\chi_\rho(k, T)$ for a fixed detuning parameter $\delta g = 1.12 \times 10^{-2}$ and a range of temperatures T. At finite temperatures the compressibility $\chi_\rho(k = 0, T)$ is non-zero. In addition, the curvature at zero momentum, $\frac{d^2\chi_\rho(k,T)}{dk^2}|_{k=0}$, decreases at finite temperatures. Both effects should be taken into account in measurements of the capacitance at finite temperature.

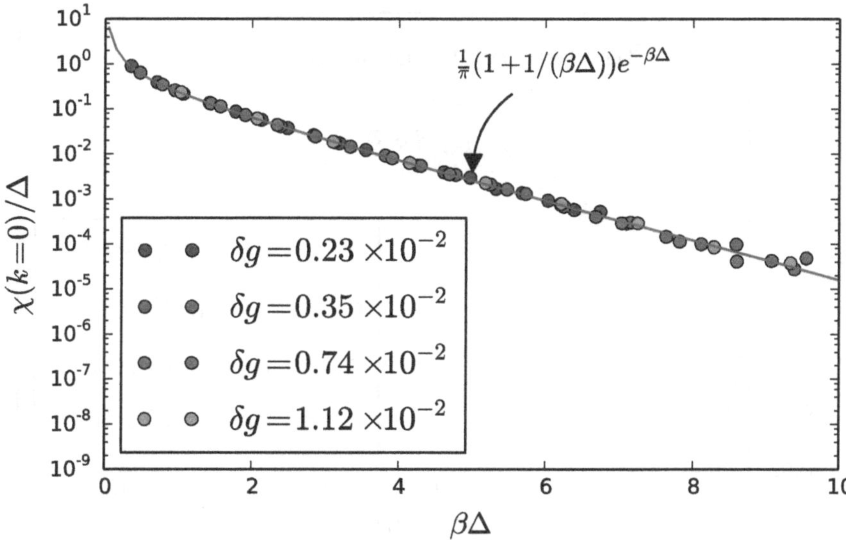

Fig. 3.4 Scaling function of the compressibility $\chi_\rho(k = 0, \delta g, T) = \Delta\ f(\beta\Delta)$. *Curves* for different values of the detuning parameter δg collapse to a single universal *curve*. The functional form of the *curve* matches the analytic calculation for a free Klein-Gordon field

The insulator is a gapped phase, therefore we expect that at low temperatures the compressibility will follow an activated behavior [18]. This is demonstrated in Fig. 3.4, where we plot the compressibility as a function of the inverse temperature β. The compressibility scales near the critical point as $\chi_\rho(k = 0, T) = \Delta\ f(\beta\Delta)$, and both axes in Fig. 3.4 are rescaled to obtain the collapsed universal scaling function f. Indeed, we find that at low temperatures, $\beta\Delta \gg 1$, the compressibility decays exponentially $\chi_\rho(k = 0, T) \sim e^{-\Delta/T}$.

We define a generalization of the capacitance to finite temperatures as the curvature of charge susceptibility at zero wave number,

$$C(T) = \frac{1}{2}\frac{\partial^2\chi_\rho(k, T)}{\partial k^2}|_{k=0}\,. \tag{3.11}$$

As $T \to 0$, this coincides with the zero temperature definition. In Fig. 3.5 we depict the generalized capacitance as a function of the inverse temperature β. As before, both axes are rescaled to obtain the scaling function. The curves rapidly converge to the zero temperature limit, allowing for an accurate determination of the capacitance for $\beta\Delta > 6$.

As a point of reference, we compare our compressibility to that of a simple analytic calculation based on a free complex Klein-Gordon boson field with mass Δ, with Euclidean action

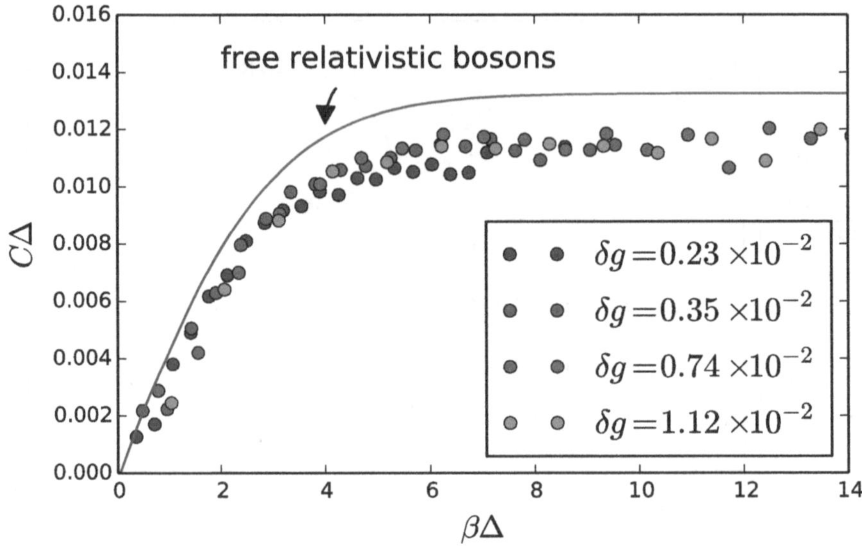

Fig. 3.5 Curvature of the charge susceptibility, $C(T) = \frac{\partial^2 \chi_\rho(k,T)}{\partial k^2}|_{k=0}$. For $T \to 0$, this becomes the capacitance. Axes are rescaled to obtain scaling behavior. The noise is dominated by numerical derivatives of the QMC data. The *solid line* shows an analytic calculation of C for a free Klein-Gordon field

$$S = \int_0^\beta d\tau \int d^2x \left[|\partial_\tau \varphi|^2 + |\nabla\varphi|^2 + \Delta^2 |\varphi|^2 \right] . \tag{3.12}$$

This gives the leading result in a $1/N$ expansion, using the renormalized mass Δ as an input. The static charge susceptibility is then obtained from a one-loop Feynman diagram calculation, [14]

$$\chi_\rho(k, \nu_m = 0, T) = \frac{1}{\beta} \sum_{\omega_m} \int \frac{d^2p}{(2\pi)^2} \left[\frac{4\omega_m^2}{\left(\omega_m^2 + p^2 + \Delta^2\right)\left(\omega_m^2 + (k+p)^2 + \Delta^2\right)} - \frac{2}{\omega_m^2 + p^2 + \Delta^2} \right] , \tag{3.13}$$

where the second term is the diamagnetic contribution. At $k = 0$ this gives the temperature dependent compressibility,

$$\chi_\rho(k = 0, \nu_m = 0, T) = \int_0^\infty \frac{p\,dp}{2\pi} \frac{\beta}{\cosh(\beta\omega_p) - 1} \sim \frac{\Delta}{\pi} \left(1 + 1/(\beta\Delta)\right) e^{-\beta\Delta} . \tag{3.14}$$

where $\omega_p = \sqrt{p^2 + \Delta^2}$, and where the last expression is asymptotic in the limit $\beta\Delta \gg 1$. This expression is shown in Fig. 3.4 to match closely the numerical data, despite the simplicity of the model. In addition, from Eq. (3.13) we compute the finite temperature curvature of the charge susceptibility, Eq. (3.11). We obtain the integral expression

$$\frac{1}{2}\frac{\partial^2 \chi_\rho}{\partial k^2}\bigg|_{k=0} = \int_0^\infty \frac{p\,dp}{96\pi\omega_p^5}\left\{\beta\omega_p \operatorname{csch}^2\left(\frac{\beta\omega_p}{2}\right)\left[6\Delta^2 + \beta^2 p^2\omega_p^2\left(3\operatorname{csch}^2\left(\frac{\beta\omega_p}{2}\right)+2\right)\right.\right.$$
$$\left.\left. -6\beta\omega_p^3 \coth\left(\frac{\beta\omega_p}{2}\right)\right] + 12\Delta^2 \coth\left(\frac{\beta\omega_p}{2}\right)\right\} \tag{3.15}$$

where $\operatorname{csch} x \equiv 1/\sinh x$. This is evaluated numerically in Fig. 3.5. In this case, the analytic result captures the qualitative temperature dependence, although it does not yield the correct overall scale.

3.7 High Frequency Non-universal Corrections to the Dynamical Conductivity

The universal scaling function of the dynamical conductivity is obtained by rescaling the Matsubara frequency axis by the single particle gap Δ. Curves for different detuning parameters δg collapse into a single universal at low frequencies. On the other hand, at high frequencies, ω_m need not be a negligible fraction of the ultra-violet (UV) cutoff scale Λ. This leads to non-universal corrections in the conductivity that depend on powers of ω_m/Λ. We take these into account by fitting the numerical QMC data to the following scaling form

$$\sigma_\pm(\omega_m, \delta g, \Lambda) = \sigma_q \Sigma_\pm\left(\frac{\omega_m}{\Delta}\right) + A\,\frac{\omega_m}{\Lambda} + B\left(\frac{\omega_m}{\Lambda}\right)^2. \tag{3.16}$$

Here, A and B are expected to depend smoothly on the detuning parameter δg. Since we consider a narrow range of values of δg, we approximate A and B as constants. This enables us to extract the universal functions $\Sigma_\pm$ on *both phases* by using only two fitting parameters.

To verify our anasatz we determined the parameters A and B by a numerical fit to Eq. 3.16 for a range of frequencies ω/Δ on both sides of the phase transition. The results of this analysis are presented in Fig. 3.6. We find that both A and B are *independent* of (ω/Δ) and do not change when crossing the SIT. This provides strong evidence for validity of our ansatz and demonstrates that the correction to scaling originates from the short range physics at the cut off scale Λ and not from the correction to the critical energy scale Δ. To increase the accuracy of our fit we performed a joint fit, using all frequency data points.

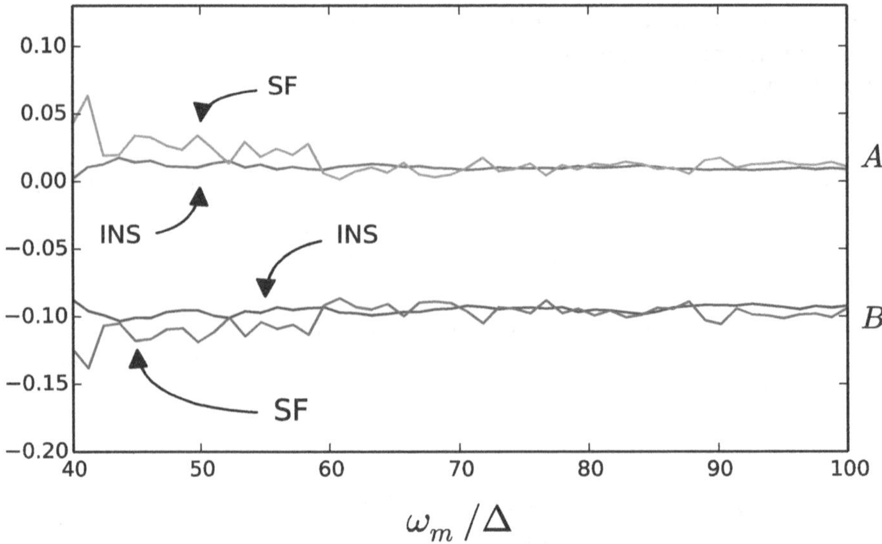

Fig. 3.6 Numerical fit of the parameters A and B in Eq. 3.16. Both quantities show little variation with respect to Δ or δg. Importantly, they vary smoothly across the phase transition, as the values of A and B do not change significantly from the insulating (INS) phase to the superfluid (SF)

3.8 Universal Scaling Function of the Matrsubara Conductivity and Charge Vortex Duality at Finite Frequency

The universal scaling function of the Matsubara conductivity is obtained by subtracting the non universal part of the conductivity using Eq. (3.16). The result is depicted in Fig. 3.7. The conductivity curves, on each side of the phase transition, collapse, with high accuracy, to the universal conductivity functions $\Sigma_{\pm}(\omega_m/\Delta)$.

At high frequencies the universal conductivity curves saturates to a plateau, with $\sigma(\omega \gg \Delta) = 0.354(5)\,\sigma_q$ in the insulating phase and $\sigma(\omega \gg \Delta) = 0.355(5)\,\sigma_q$ in the superfluid phase. As a result, we conclude that the high frequency universal conductivity, σ^*, is a robust quantity across the phase transition. Our scaling correction analysis differs significantly from that of Refs. [9, 10], yet the value of the high-frequency conductivity is in agreement with their results.

Finally, we study the CVD as a function of Matsubara frequency. In Fig. 3.8 we depict the product of the Matsubara frequency conductivity evaluated at mirror points across the critical point, $\mathcal{R}(\omega_m) = \sigma(\omega_m, \delta g)\sigma(\omega_m, -\delta g)/\sigma_q^2$. In order to study the critical properties we subtract the non-universal cut-off corrections. Note that for $\omega_m \gg \Delta$, $\mathcal{R} \rightarrow (\sigma^*/\sigma_q)^2$, whereas for $\omega_m \ll \Delta$, $\mathcal{R}$ approaches the product of reactive conductivities in the two phases. In both limits, the Matsubara and real frequency results coincide, $\mathcal{R}(\omega) = \mathcal{R}(\omega_m)$. By contrast, at intermediate frequencies, determination of $\mathcal{R}(\omega)$ requires analytical continuation. If the CVD were exact then

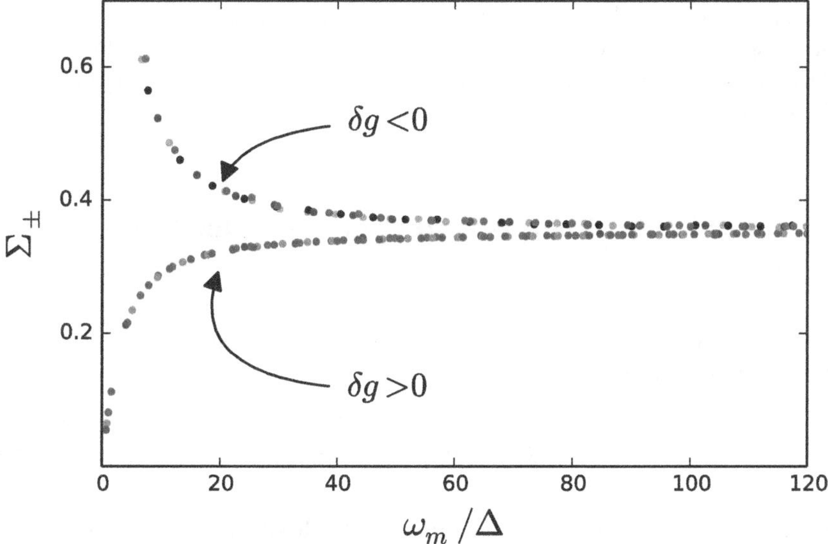

Fig. 3.7 Scaling function of the dynamical conductivity in the superfluid ($\delta g < 0$) and insulator ($\delta g > 0$). Data for different values of the detuning δg collapse to two universal *curves*

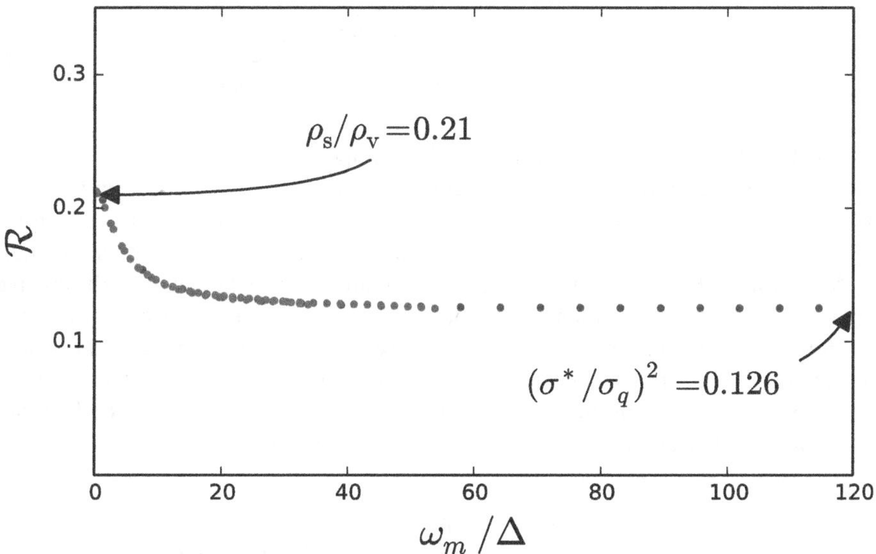

Fig. 3.8 Measure of charge vortex duality of the $O(2)$ model. Universal scaling function for $\mathcal{R}(\omega_m)$ defined in Eq. (3.2). Deviation of this function from unity quantifies the difference between charge and vortex matter

Eq. (3.1) would imply that this product is frequency independent and equal to 1. Our results display a non trivial frequency dependence and deviate from the predicted CVD value. We attribute this deviation to the different interaction range of charges and vortices.

3.9 Implications of Charge-Vortex Duality on the Higgs Mass

Assuming that CVD is exact yields a relation between the optical conductivities on both sides of the transition:

$$\sigma(\omega, -\delta g) = \frac{\sigma_{\mathrm{q}}^2}{\sigma(\omega, \delta g)} . \tag{3.17}$$

Here, $\sigma(\omega, \delta g)$ is complex, containing both dissipative and reactive parts $\sigma = \sigma' + i\sigma''$, such that

$$\sigma'(\omega, -\delta g) = \frac{\sigma_{\mathrm{q}}^2 \sigma'(\omega, \delta g)}{\sigma'^2(\omega, \delta g) + \sigma''^2(\omega, \delta g)} \tag{3.18}$$

$$\sigma''(\omega, -\delta g) = -\frac{\sigma_{\mathrm{q}}^2 \sigma''(\omega, \delta g)}{\sigma'^2(\omega, \delta g) + \sigma''^2(\omega, \delta g)} \tag{3.19}$$

Note that duality flips the sign of the reactive component.

Numerically we found it difficult to extract the reactive part of the conductivity. The results for the analytic continuation were much less numerically stable than for the dissipative component. Yet, the numerics do provide some evidence for the duality. According to Eq. (3.18) one prediction of duality is that whenever the dissipative part vanishes for some frequency ω in one of the phases, it must also vanish at the mirror point in the other phase. This is indeed seen to be the case in Fig. 2.11, where the threshold frequency of the dissipative part of the optical conductivity equals $\omega_{\mathrm{T}} \sim 2\Delta$ on *both* sides of the transition. The presence of small subgap conductivity in the superfluid is a consequence of the inexactness of the duality.

3.10 Charge-Vortex Duality at One Loop Order in Weak Coupling

As an additional test of the duality, in Appendix C we present analytic calculations of the optical conductivity on both sides of the transition, to one loop order. In Figs. 3.9 and 3.10 we show the dynamical conductivity on the ordered and disordered phase,

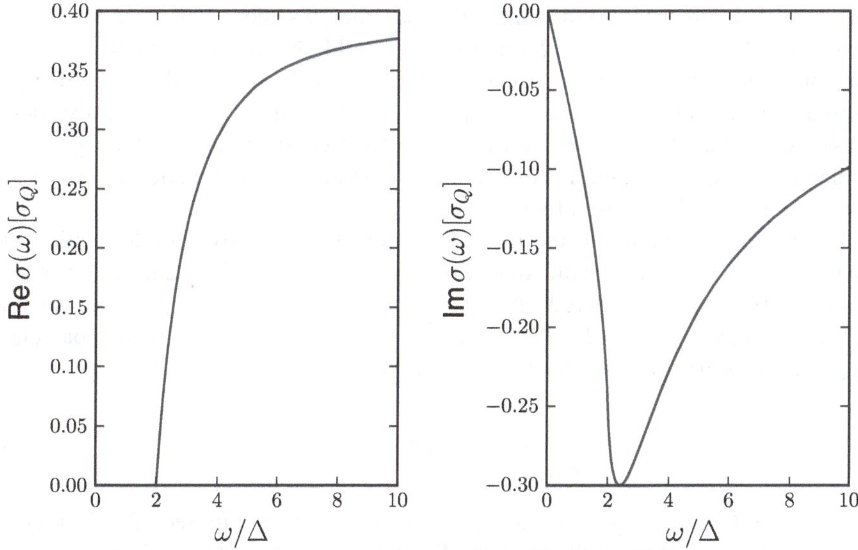

Fig. 3.9 The real and imaginary part of the optical conductivity in the disordered phase. Results are shown from a one loop calculation in Appendix C

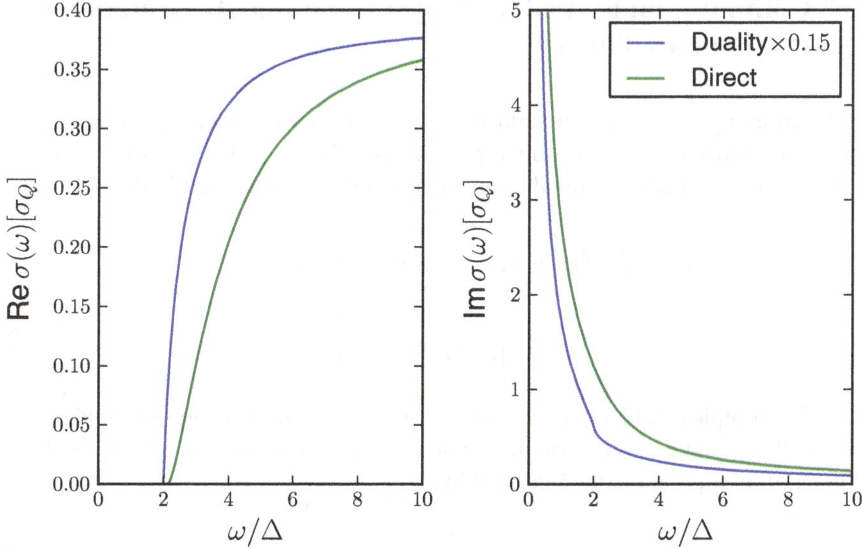

Fig. 3.10 The real and imaginary part of the optical conductivity in the ordered phase. The *green curve* displays the results of a one loop calculation carried in Appendix C. The values of m_H/Δ and ρ_s/Δ were taken from the QMC simulation. The *blue curve* depicts the optical conductivity obtained from the duality relation in Eq. (3.17). This curve is multiplied by 0.15 for comparison reasons

respectively. In order to use the same reference energy scale in both figures, we used the universal values $m_H/\Delta = 2.1$ and $\rho_s/\Delta = 0.44$ obtained numerically in earlier parts of the analysis. In Fig. 3.10 we also depict the conductivity in the ordered phase, as obtained by applying the duality, Eq. (3.17), to the conductivity in the disordered phase. As in the DC case, the overall scale of the conductivity has the right order of magnitude, set by σ_q, but is not quantitative. However, the functional form of the conductivity is well captured by the duality.

In addition, we compute the universal ratio of the reactive conductivities $\frac{C_{\text{ins}}}{L_{\text{sf}}}$ from the weak coupling result. Indeed we find that at low frequencies the optical conductivity in the disordered phase, computed in Appendix C, rises linearly as $\sigma''_{\text{dis}}(\omega) = -2\pi\sigma_q \times \hbar\omega/(24\pi\Delta) + \mathcal{O}(\omega^2)$, that is, as a capacitor with capacitance $C_{\text{ins}} = 2\pi\sigma_q \times \hbar/(24\pi\Delta)$. This yields the ratio

$$\frac{C_{\text{ins}}}{L_{\text{sf}}} = \frac{2\pi\rho_s}{12\Delta}\sigma_q^2 \approx 0.23\,\sigma_q^2\,. \tag{3.20}$$

where in the last equality we used $\rho_s/\Delta = 0.44$ as obtained in Sect. 2.4.3. Interestingly, this value is close to the numerically exact result $C_{\text{ins}}/L_{\text{sf}} = 0.21\sigma_q^2$.

3.11 Computation of the Dynamical Conductivity from the Dual Vortex Theory

An illuminating way to understand the low frequency conductivity is through the dual vortex representation. In this representation the effective field theory is given by a complex ψ^4 theory coupled to an electromagnetic gauge field [19, 20]:

$$\begin{aligned} S = \int d^3x \Big\{ &|(\partial_\mu + i a_\mu)\psi|^2 + m^2|\psi|^2 + \lambda|\psi|^4 \\ &+ \frac{1}{16\pi^2 K} f_{\mu\nu} f_{\mu\nu} \Big\}\,. \end{aligned} \tag{3.21}$$

Here, the complex field ψ is the vortex condensate order parameter field, $f_{\mu\nu} = \partial_\mu a_\nu - \partial_\nu a_\mu$, and K is the coupling constant of the bosons. The gauge field a_μ is related to the original boson 3-current by:

$$J_\mu = \frac{1}{2\pi}\epsilon_{\mu\nu\lambda}\partial_\nu a_\lambda \tag{3.22}$$

Since the current is equal to the dual electric field $J_x(i\omega_m) = -\omega_m a_y/2\pi$, the conductivity is

$$\begin{aligned}\sigma(i\omega_m) &= -\frac{1}{\omega_m}\big\langle J_x(i\omega_m)\, J_x(-i\omega_m)\big\rangle \\ &= \frac{\omega_m}{(2\pi)^2}\big\langle a_y(i\omega_m)\, a_y(-i\omega_m)\big\rangle.\end{aligned} \tag{3.23}$$

In the disordered vortex phase, corresponding to the superfluid phase of the original bosons, the gauge field remains gapless with the propagator in Feynman gauge:

$$\big\langle a_\mu\, a_\nu\big\rangle = \frac{4\pi^2 K}{k^2}\,\delta_{\mu\nu}\,, \tag{3.24}$$

hence the conductivity is $\sigma_{\rm sf}(i\omega_m) = K/\omega_m$. After analytic continuation and introducing physical units $e^{*2}/\hbar = 2\pi\sigma_{\rm q}$, this becomes

$$\sigma_{\rm sf}(\omega) = 2\pi\sigma_{\rm q} \times \frac{\rho_s}{\hbar}\left[\frac{i}{\omega} + \pi\delta(\omega)\right], \tag{3.25}$$

where we have set $K = \rho_s$, its value in the superfluid phase. In the condensed vortex state, corresponding to the disordered phase of the original bosons, the field ψ gets an expectation value leading to a mass term for the gauge field through the Anderson-Higgs mechanism. The effective action of the gauge field is then given by a Proca action:

$$\mathcal{S} = \int d^3x\left\{\frac{1}{16\pi^2 K}\, f_{\mu\nu} f_{\mu\nu} + \frac{1}{2}\rho_v a_\mu^2\right\}. \tag{3.26}$$

where we now take the vortex condensation density $\rho_v = 2|\langle\psi\rangle|^2$. The gauge field propagator is

$$\big\langle a_\mu\, a_\nu\big\rangle = \frac{4\pi^2 K}{k^2 + M^2}\left(\delta_{\mu\nu} - \frac{k_\mu k_\nu}{M^2}\right), \tag{3.27}$$

where the gauge field mass M is given by $M^2 = 4\pi^2 K\rho_v$. Note that, since the current is quadratic in boson operators, this mass is related to the single-boson gap Δ by $M = 2\Delta$. Now the conductivity is given by $\sigma_{\rm dis}(i\omega_m) = \omega_m K/(\omega_m^2 + M^2)$, which yields after analytic continuation $\sigma_{\rm dis}(\omega) = i\omega K/(\omega^2 - M^2)$. At low frequencies $\omega \ll M$ this becomes, in physical units,

$$\sigma_{\rm ins}(\omega) \approx -2\pi i\sigma_{\rm q}\frac{K}{M^2}\,\hbar\omega = -i\sigma_{\rm q}\frac{\hbar\omega}{2\pi\rho_v}. \tag{3.28}$$

Combining the results from Eqs. (3.25) and (3.28) we obtain

$$\frac{C_{\rm ins}}{L_{\rm sf}} = \frac{\rho_s}{\rho_v}\sigma_{\rm q}^2. \tag{3.29}$$

This reestablishes a physical interpretation for the universal ratio $C_{\text{ins}}/L_{\text{sf}}$ as the ratio between the superfluid stiffness and the vortex condensation density on opposite sides of the transition.

3.12 Summary

The universal ratio of the reactive conductivity $C_{\text{ins}}/L_{\text{sf}}$ motivates future experiments as it provides a direct probe of the charge-vortex duality.

Recent THz spectroscopy measured the complex AC conductivity near the SIT in superconducting InO and NbN thin films [21]. In these systems, the superfluid stiffness in the superconducting phase can be measured from the low frequency reactive response [22, 23]. Detecting the diverging capacitance in the insulator side may require careful subtraction of substrate signal background [24].

Another experimental realization of the SIT is the Mott insulator to superfluid transition of cold atom trapped in an optical lattice. We propose a direct approach to extract the capacitance of the Mott insulator using *static* measurement of the charge susceptibility. This can be measured, e.g. by applying an optical potential at small wave vector k and probing the rearrangement of boson density using in-situ imaging [25].

Alternatively $\sigma'(\omega)$, for which experimental protocols were proposed, [9, 26] can be used to compute $\sigma''(\omega)$ by means of the Kramers-Kronig integral.

In summary, we computed the vortex condensate stiffness ρ_{v}, the high frequency universal conductivity and provided a quantitative measure for deviation from CVD as a function of Matsubara frequency. In addition, we suggested concrete experiments that test our predictions in Thz spectroscopy of thin superconducting films and in cold atomic systems.

References

1. M.P.A. Fisher, D.H. Lee, Correspondence between two-dimensional bosons and a bulk superconductor in a magnetic field. Phys. Rev. B **39**(4), 2756–2759 (1989). doi:10.1103/PhysRevB.39.2756. http://link.aps.org/doi/10.1103/PhysRevB.39.2756
2. M.P.A. Fisher, G. Grinstein, S.M. Girvin, Presence of quantum di usion in two dimensions: Universal resistance at the superconductor-insulator transition. Phys. Rev. Lett. **64**(5), 587–590 (1990). doi:10.1103/PhysRevLett.64.587. http://link.aps.org/doi/10.1103/PhysRevLett.64.587
3. D. Shahar et al., Evidence for charge-flux duality near the quantum hall liquidto-insulator transition. Science **274**(5287), 589–592 (1996). doi:10.1126/science.274.5287.589. http://www.sciencemag.org/content/274/5287/589.full.pdf. http://www.sciencemag.org/content/274/5287/589.abstract
4. A.M. Goldman, Superconductor-insulator transitions. Int. J. Mod. Phys. B 24.20n21, 4081–4101 (2010). doi:10.1142/S0217979210056451. http://www.worldscientific.com/doi/abs/10.1142/S0217979210056451

5. M.E. Peskin, Mandelstam't Hooft duality in abelian lattice models. Ann. Phys. **113**(1), 122–152 (1978). issn: 0003–4916. doi:10.1016/0003-4916(78)90252-X. http://www.sciencedirect.com/science/article/pii/000349167890252X
6. C. Dasgupta, B.I. Halperin, Phase transition in a lattice model of superconductivity. Phys. Rev. Lett. **47**(21), 1556–1560 (1981). doi:10.1103/PhysRevLett.47.1556. http://link.aps.org/doi/101103/PhysRevLett.47.1556
7. S. Gazit, D. Podolsky, A. Auerbach, Fate of the higgs mode near quantum criticality. Phys. Rev. Lett. **110**(14), 140401 (2013). doi:10.1103/PhysRevLett.110.140401. http://link.aps.org/doi/10.1103/PhysRevLett.110.140401
8. S. Gazit et al., Dynamics and conductivity near quantum criticality. Phys. Rev. B **88**(23), 235108 (2013). doi:10.1103/PhysRevB.88.235108. http://link.aps.org/doi/10.1103/PhysRevB.88.235108
9. K. Chen et al., Universal conductivity in a two-dimensional super fluid-to-insulator quantum critical system. Phys. Rev. Lett. **112**(3), 030402 (2014). doi:10.1103/PhysRevLett.112.030402. http://link.aps.org/doi/10.1103/PhysRevLett.112.030402
10. W. Witczak-Krempa, E.S. Sorensen, S. Sachdev, The dynamics of quantum criticality revealed by quantum Monte Carlo and holography. Nat. Phys. **10**(5), 361–366 (2014). ISSN: 1745–2473. doi:10.1038/nphys2913
11. A. Auerbach, D.P. Arovas, S. Ghosh, Quantum tunneling of vortices in two-dimensional condensates. Phys. Rev. B **74**(6), 064511 (2006). doi:10.1103/PhysRevB.74.064511. http://link.aps.org/doi/10.1103/PhysRevB.74.064511.(Bibliography 79)
12. D. Podolsky, S. Sachdev, Spectral functions of the Higgs mode near two-dimensional quantum critical points. Phys. Rev. B **86**(5), 054508 (2012). doi:10.1103/PhysRevB.86.054508. http://link.aps.org/doi/10.1103/PhysRevB.86.054508
13. D. Podolsky, A. Auerbach, D.P. Arovas, Visibility of the amplitude (Higgs) mode in condensed matter. Phys. Rev. B **84**(17), 174522 (2011). doi:10.1103/PhysRevB.84.174522. http://link.aps.org/doi/10.1103/PhysRevB.84.174522
14. K. Damle, S. Sachdev, Nonzero-temperature transport near quantum critical points. Phys. Rev. B **56**(14), 8714–8733 (1997). doi:10.1103/PhysRevB.56.8714. http://link.aps.org/doi/10.1103/PhysRevB.56.8714
15. M. Hasenbusch, T. Torok, High-precision Monte Carlo study of the 3D XY-universality class. J. Phys. A: Math. Gen. **32**(36), 6361–6371 (1999). issn: 0305–4470, 1361–6447, doi:10.1088/0305-4470/32/36/301. http://iopscience.iop.org/0305-4470/32/36/301 Accessed 18 June 2012
16. E.L. Pollock, D.M. Ceperley, Path-integral computation of super uid densities. Phys. Rev. B **36**(16), 8343–8352 (1987). doi:10.1103/PhysRevB.36.8343. http://link.aps.org/doi/10.1103/PhysRevB.36.8343
17. D.J. Scalapino, S.R. White, S. Zhang, Insulator, metal, or superconductor: the criteria. Phys. Rev. B **47**(13), 7995–8007 (1993). doi:10.1103/PhysRevB.47.7995. http://link.aps.org/doi/10.1103/PhysRevB.47.7995
18. M. Endres, Probing correlated quantum many-body, systems at the single-particle level. PhD thesis. (Ludwig-Maximilians-Universitaat Munchen, 2013)
19. M. Stone, P.R. Thomas, Condensed monopoles and abelian confinement. Phys. Rev. Lett. **41**(6), 351–353 (1978). doi:10.1103/PhysRevLett.41.351. http://link.aps.org/doi/10.1103/PhysRevLett.41.351
20. D.P. Arovas, J.A. Freire, Dynamical vortices in super uid films. Phys. Rev. B **55**(2), 1068–1080 (1997). doi:10.1103/PhysRevB.55.1068. http://link.aps.org/doi/10.1103/PhysRevB.55.1068
21. D. Sherman et al., E ect of Coulomb interactions on the disorder-driven superconductor-insulator transition. Phys. Rev. B **89**(3), 035149 (2014). doi:10.1103/PhysRevB.89.035149. http://link.aps.org/doi/10.1103/PhysRevB.89.035149
22. J. Corson et al., Vanishing of phase coherence in underdoped Bi2Sr2CaCu2O8+. En. Nature **398**(6724), 221–223 (1999). issn: 0028–0836. doi:10.1038/18402. http://www.nature.com/nature/journal/v398/n6724/full/398221a0.html. Accessed 28 Apr 2014
23. R.W. Crane et al., Fluctuations, dissipation, and nonuniversal super uid jumps in two-dimensional superconductors. Phys. Rev. B **75**(9), 094506 (2007). doi:10.1103/PhysRevB.75.094506. http://link.aps.org/doi/10.1103/PhysRevB.75.094506. Accessed 28 Apr 2014

24. A. Frydman, Private communication. 2014
25. J.F. Sherson et al., Single-atom-resolved uorescence imaging of an atomic Mott insulator. Nature 467.7311, 68–72 (2010). issn, 0028–0836, doi:10.1038/nature09378
26. A. Tokuno, T. Giamarchi, Spectroscopy for cold atom gases in periodically phase-modulated optical lattices. Phys. Rev. Lett. **106**(20), 205301 (2011). doi:10.1103/PhysRevLett.106.205301. http://link.aps.org/doi/10.1103/PhysRevLett.106.205301

Chapter 4
Summary and Outlook

In this thesis we explored a few aspects of dynamics near quantum criticality in two space dimensions. More specifically, we studied the universal properties of the amplitude (Higgs) mode in the ordered phase of relativistic $O(N)$ models and the charge-vortex duality near the SIT.

The universal scaling function of the scalar susceptibility and the optical conductivity were computed by means of a large scale QMC simulation combined with a numerical analytic continuation. It was shown that the amplitude mode can be probed arbitrarily close to the critical point in the spectrum of the foregoing response functions. In particular, the scalar susceptibility contains a well defined resonance that can be associated with the amplitude mode and the optical conductivity has a sharp threshold at the amplitude mode resonance frequency. The universal amplitude ratio between the amplitude mode mass and the single particle gap was determined to be $m_H/\Delta \approx 2.1$.

The critical capacitance in the insulating phase was identified as a measure of the vortex condensate stiffness ρ_{v}. The capacitance was computed directly from a QMC simulation without the use of a numerical analytic continuation. We determined the amplitude ratio between the superfluid stiffness and the vortex condensate stiffness to be $\rho_{\mathrm{s}}/\rho_{\mathrm{v}} \approx 0.21\sigma_{\mathrm{q}}^2$. This value deviates from the charge-vortex duality prediction of $\rho_{\mathrm{s}}/\rho_{\mathrm{v}} = \sigma_{\mathrm{q}}^2$ thus it provides a quantitative measure of CVD violation. In addition, we calculated the high frequency universal conductivity, $\sigma^* \approx 0.35\sigma_{\mathrm{q}}$. Finally, we suggest a concrete cold atomic experiment for measuring the capacitance by in-situ imaging techniques.

As technical progress, we present a novel QMC method based on the worm algorithm for simulating $O(N)$ models and a technique for analyzing the errors of numerical analytic continuation based on singular value decomposition.

Given the universal nature of our results, they motivate future experiments that probe dynamical properties near quantum criticality, such as in disordered superconductors and cold atomic systems.

An intriguing extension to this work, would be to study the effect of varying degrees of disorder on our results. This is especially important since disorder is

S. Gazit, *Dynamics Near Quantum Criticality in Two Space Dimensions*,
Springer Theses, DOI 10.1007/978-3-319-19354-0_4

present in any realistic experimental setting and in certain cases it drives the phase transition. It would be interesting to study the visibility of the amplitude mode and to further test the charge vortex duality in the presence of disorder.

Another interesting research direction would be to study the signature of the amplitude mode in one dimensional quantum systems. This in particularly interesting since according to the Mermin-Wagner theorem the strong fluctuations destroy the long range order leaving only a power-law quasi-long range order.

Appendix A
Worm Algorithm for $O(N)$ Models

We present a novel QMC algorithm for O(N) lattice models Eq. (2.9). The algorithm is based on the worm algorithm [1] extending it for general O($N > 2$) models. The first step is to expand Eq. (2.9) in strong coupling:

$$\mathcal{Z} = \int \mathcal{D}\vec{\phi} \prod_{b} \prod_{\alpha} \sum_{n_b^\alpha} \frac{1}{n_b^\alpha!} (\phi_{i_b}^\alpha \phi_{i'_b}^\alpha)^{n_b^\alpha} \prod_j e^{-V(|\vec{\phi}_j|^2)} \tag{A.1}$$

with $\mathcal{D}\vec{\phi} \equiv \prod_i d^N\phi_i$. Here $\{b\}$ represent the set of all lattice bonds, the site i_b is linked to the site i'_b through the bond b, the index $\alpha \in \{1, \ldots, N\}$ labels the N components of each $\vec{\phi}_i$, and $V(s) = \mu s + g s^2$ is the local on-site interaction. Next we integrate out the fields $\vec{\phi}_i$. This can be achieved by noting that now the functional integral factorizes into a product of *single site integrals*, such that

$$\mathcal{Z} = \sum_{\{n_b^\alpha\}} \prod_{b,\alpha} \frac{1}{n_b^\alpha!} \prod_i W\left(\{k_i^\alpha\}\right) . \tag{A.2}$$

where we define $k_i^\alpha = \sum'_{b(i)} n_b^\alpha$ as the sum over all bonds b emanating from site i. The single site weight is then

$$W\left(\{k_i^\alpha\}\right) = \int d^N\phi_i \prod_\alpha \langle \phi_i^\alpha |^{k_i^\alpha} e^{-V(|\vec{\phi}|^2)} . \tag{A.3}$$

We may write

$$\begin{aligned} W\left(\{k_i^\alpha\}\right) &= \int d^N\phi_i \int_0^\infty ds\, e^{-V(s)}\, \delta\left(s - |\vec{\phi}_i|^2\right) \prod_\alpha (\phi_i^\alpha)^{k_i^\alpha} \\ &= \frac{1}{2\pi} \int_0^\infty ds\, e^{-V(s)} \int_{-\infty}^\infty d\lambda\, e^{i\lambda s} \prod_\alpha \mathcal{I}(k_i^\alpha) , \end{aligned} \tag{A.4}$$

S. Gazit, *Dynamics Near Quantum Criticality in Two Space Dimensions*, Springer Theses, DOI 10.1007/978-3-319-19354-0

where

$$\mathcal{I}(k_i^\alpha) = \int_{-\infty}^{\infty} d\phi_i^\alpha \, e^{-i\lambda(\phi_i^\alpha)^2} \, (\phi_i^\alpha)^{k_i^\alpha} = (i\lambda)^{-(k_i^\alpha+1)/2} \, \Gamma\big(\frac{1}{2} + \frac{1}{2}k_i^\alpha\big) \, \delta_{k_i^\alpha,\text{even}} \, . \tag{A.5}$$

We now encounter the integral

$$\int_{-\infty}^{\infty} d\lambda \, e^{i\lambda s} \, (i\lambda)^{-J} = 2\, s^{J-1} \, \Gamma\big(1 - J\big) \, \sin(\pi J) \, , \tag{A.6}$$

where $J = \frac{1}{2}(N + K_i)$, and $K_i = \sum_\alpha k_i^\alpha$. The above integral converges only if $0 < \text{Re}J < 1$, however our initial expression in Eq. A.3 is clearly convergent for all possible values of J, which licenses us to analytically continue the above expression, using the identity $\Gamma(J)\,\Gamma(1-J) = \pi/\sin(\pi J)$. We then obtain

$$W\big(\{k_i^\alpha\}\big) = Q\big(\frac{1}{2}N + \frac{1}{2}K_i\big) \prod_\alpha \Gamma\big(\frac{1}{2} + \frac{1}{2}k_i^\alpha\big) \, \delta_{k_i^\alpha,\text{even}} \, , \tag{A.7}$$

with

$$Q(J) = \frac{1}{\Gamma(J)} \int_0^\infty ds \, e^{-V(s)} \, s^{J-1} \, . \tag{A.8}$$

The one-dimensional integrals $Q(J)$ can be evaluated numerically to high precision and tabulated prior to the QMC simulation. In this representation the partition function sum runs over all integer values of the bond's strength n_b^α, replacing the $\vec{\phi}_i$ field integrations. The sum is restricted only to closed path loops due to constraint $\delta_{k_i^\alpha,\text{even}}$.

The updating procedure closely follows the worm algorithm, considering an extended partition function:

$$\mathcal{Z}_G = \sum_{i,j} \langle \phi_i^\alpha \phi_j^\alpha \rangle \tag{A.9}$$

The fields insertion $\phi_i^\alpha \phi_j^\alpha$ breaks the closed path condition by adding a single open loop. The open loop's head is located at i and its tail at j.

For simplicity we choose the open loop to be one of the flavors α. The updating procedure consists out of two elementary steps. The first move is a shift move in which we move the worm's head to one of the neighboring sites connected with the bond b. During the move we either increase or decrease the bond's strength n_b^α. The second move is a jump move, which is relevant only for closed loops where the head and the tail are located in the same site. We choose one of the lattice sites and jump with the head tail pair to that site. The QMC acceptance ratios can be easily derived from Eqs. (A.7) and (A.2) similarly to the argument in Ref. [1].

We tested the correctness of our numerical implementation by comparing with previous QMC simulation and to analytic results of the Gaussian model limit of Eq. 2.9 ($g = 0$). The results agree within the statistical errors.

We also provide an explicit expression for the sampling of the scalar susceptibility in the closed path representation. The operator insertion $\vec{\phi}_i^2$ effectively introduces a factor of s to the integrand in Eq. A.4, in which case Eq. A.8 is replaced by $J_i\ Q(J_i + 1)$. Inserting $(\vec{\phi}_i^2)^2$ introduces a factor of s^2 and results in $J_i(J_i+1)\ Q(J_i+2)$. Thus, the insertion $\vec{\phi}_i^2\vec{\phi}_j^2$ yields

$$\begin{aligned}\langle\vec{\phi}_i^2\vec{\phi}_j^2\rangle &= \left\langle\frac{J_i J_j\ Q(J_i+1)\ Q(J_j+1)}{Q(J_i)\ Q(J_j)}\right\rangle \quad (i \neq j)\\ &= \left\langle\frac{J_i(J_i+1)\ Q(J_i+2)}{Q(J_i)}\right\rangle \qquad (i = j)\end{aligned} \tag{A.10}$$

Reference

1. N. Prokof'ev, B. Svistunov, Worm algorithms for classical statistical models. Phys. Rev. Lett. **87**(16), 160601 (2001). doi:10.1103/PhysRevLett.87.160601. http://link.aps.org/doi/10.1103/PhysRevLett.87.160601. Accessed 26 Mar 2012

Appendix B
Analytic Continuation of Imaginary Time Quantum Monte Carlo Data

B.1 General Formulation

We use imaginary time action Eq. 2.9 in the QMC simulations in order for the QMC weights to be real and positive, avoiding the dynamical sign problem. The real frequency, dissipative response function $A(\nu)$, can be obtained by numerical analytic continuation [1], which amounts to inverting the equation,

$$\mathcal{G}(i\omega_m) = \int_0^\infty \frac{d\nu}{\pi} \frac{2\nu}{\omega_m^2 + \nu^2} A(\nu) \,. \tag{B.1}$$

The kernel

$$K(m, \nu) = \frac{1}{\pi} \cdot \frac{2\nu}{\omega_m^2 + \nu^2}, \tag{B.2}$$

needs to be inverted in order to formally obtain,

$$A(\nu) = K^{-1}\mathcal{G}(i\omega_m) \,. \tag{B.3}$$

Unfortunately K is an ill conditioned operator. The inversion is extremely sensitive to inevitable statistical noise in $\mathcal{G}$.

The stability of the inversion problem can be analyzed by the Singular Value Decomposition (SVD)

$$K = UWV^T \tag{B.4}$$

where U and V are unitary matrices whose rows are the eigenvectors $\langle u_n|$, and $\langle v_n|$. The first five eigenvectors $v_n(\nu)$ are plotted in Fig. B.1. W is diagonal with real, non-negative SVD eigenvalues w_n. These are plotted on a logarithmic scale as a function of n in Fig. B.2. W has up to $\mathcal{N}$ non zero singular values, where $\mathcal{N}$ is the number of QMC data points.

S. Gazit, *Dynamics Near Quantum Criticality in Two Space Dimensions*,
Springer Theses, DOI 10.1007/978-3-319-19354-0

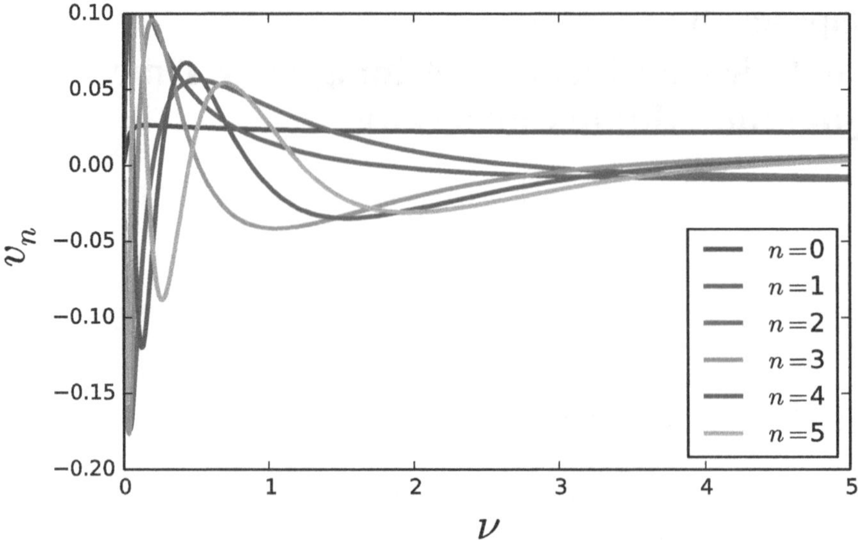

Fig. B.1 The first five vectors $v_n(\nu)$ corresponding to the largest singular values in W

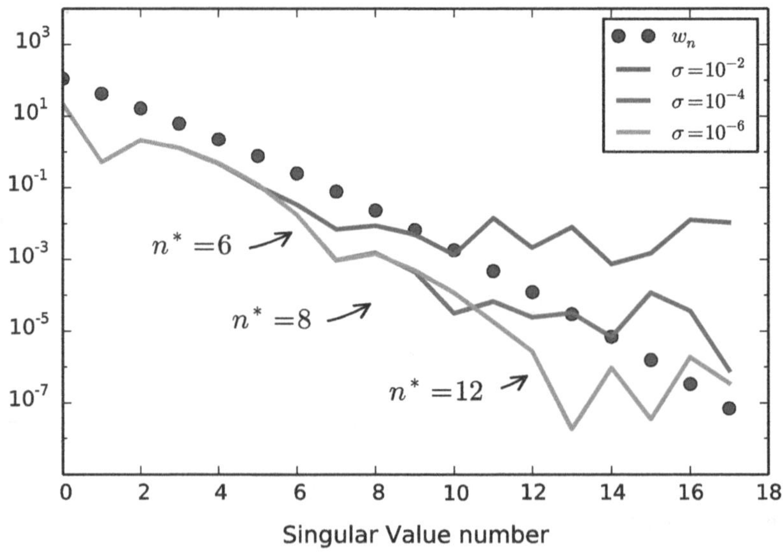

Fig. B.2 SVD analysis of the numerical analytical continuation. n labels the SVD eigenmodes. The filled circles are the rapidly decreasing SVD eigenvalues of K, denoted w_n. Magnitudes of projections of noisy data, for the test model, Eq. (B.13), are denoted by $|\tilde{p}_n|$. σ is the variance of the artificial noise added to the Matsubara data. The breakpoints n^* denotes the mode index where noise dominates the signal, and the projections start to flatten. The values of n^*increase when the noise level decreases

From Eq. (B.4), the pseudo inversion of K is given by

$$\bar{K}^{-1} = V\bar{W}^{-1}U^T. \tag{B.5}$$

Here $\bar{W}$ is a square diagonal matrix which contains only the non zero eigenvalues $w_n \neq 0$.

The SVD eigenvalues w_n can be calculated by diagonalizing the Hermitian matrix $(KK^\dagger)_{ij}$:

$$(KK^\dagger)_{ij} = \int_{-\infty}^{\infty} \frac{d\omega}{2\pi} \frac{\omega^2}{(\omega_i^2+\omega^2)(\omega_j^2+\omega^2)}$$
$$= \frac{1}{2\big(|\omega_i|+|\omega_j|\big)} = \frac{\beta}{4\pi}\frac{1}{|i|+|j|}. \tag{B.6}$$

Since $\mathcal{G}(\tau)$ is real, $\mathcal{G}_n = \mathcal{G}_{-n}$, and by projecting out the zero mode, we may restrict both i and j to be positive integers in Eq. (B.6).

Matrices of the form

$$H_{ij}(\tau,\theta) = \frac{\tau^{i+j}}{i+j+\theta}, \quad (i,j) \in \{0,\ldots,N\}. \tag{B.7}$$

are known as Hilbert matrices. We are interested in the case of $\tau = 1$ and $\theta = 2$. An exact bound on the dependence of the smallest eigenvalue on the matrix size was obtained by Ref. [2],

$$w_{\min}^{(N)} \sim \kappa\sqrt{N}\left(1+\sqrt{2}\right)^{-4N} \times \big(1+o(1)\big),$$
$$\ln\big(w_{\min}^{(N)}\big) \sim -3.52549\,N + 0.5\ln N + 0.7909, \tag{B.8}$$

with

$$\kappa = \frac{2^{15/4}\,\pi^{3/2}}{\left(1+\sqrt{2}\right)^4} = 2.205385\ldots \tag{B.9}$$

As we see, the minimal eigenvalue decreases faster than exponentially with N, which is consistent with the behavior found numerically in Fig. B.2.

B.2 Pseudo-inversion by Truncated Singular Value Decomposition

In practice, the noisy QMC data, called $\tilde{\mathcal{G}}$ can be decomposed as

$$\tilde{\mathcal{G}} = \mathcal{G}^{\mathrm{sig}} + \xi \tag{B.10}$$

where $\mathcal{G}^{\text{sig}}$ is the true signal, and ξ is a random noise. The noise interferes with the numerical inversion of $\mathcal{G}^{\text{sig}}$. To see this, the data $\tilde{\mathcal{G}}$ is projected onto the eigenvectors u_n, which yields the real numbers

$$\begin{aligned} \tilde{p}_n = \langle\, u_n \,|\, \tilde{\mathcal{G}} \,\rangle &= \langle\, u_n \,|\, \mathcal{G}^{\text{sig}} \,\rangle + \langle\, u_n \,|\, \xi \,\rangle \\ &\equiv p_n^{\text{sig}} + \xi_n. \end{aligned} \tag{B.11}$$

The pseudo inversion Eq. (B.5) yields

$$A(\nu) = \sum_n \frac{\tilde{p}_n}{w_n} v_n(\nu) = \sum_n \left(\frac{p_n^{\text{sig}}}{w_n} v_n(\nu) + \frac{\xi_n}{w_n} v_n(\nu) \right). \tag{B.12}$$

Since $\mathcal{G}^{\text{sig}}$ is the analytic continuation of a normalizable function, $\sum_n |p_n^{\text{sig}}/w_n|^2$ must converge. This implies that $|p_n^{\text{sig}}| < w_n$ at large n. On the other hand ξ_n is not the analytic continuation of a normalizable function, and therefore is not necessarily bounded by w_n. For white noise, ξ_n are random numbers whose variance is independent on n.

Therefore, one can readily identify a *breakpoint*, n^*, which for $n < n^*$, $\tilde{p}_n \approx p_n^{\text{sig}}$, and for $n > n^*$, $\tilde{p}_n \approx \xi_n$. The breakpoint serves to truncate the inversion and eliminate the dominance of noise terms. It can also allow an estimate of the truncation error.

Let us illustrate this procedure by a test model,

$$A^{\text{model}}(\nu) = \nu^3 \left(e^{-(\nu-\Delta)^2} + e^{-(\nu+\Delta)^2} \right). \tag{B.13}$$

If Fig. B.2, Eq. (B.13) to the u_n basis. w_n and $|p_n^{\text{sig}}|$ rapidly decay, as expected from the Riemann-Lebesgue lemma for a smooth spectral function. We add an artificial white noise with increasing variance σ. As expected, the (approximately) exponential decay of $\tilde{p}_n$ stops abruptly at n^*, where $|\tilde{p}_{n^*}| \approx |\xi_{n^*}|$.

As seen in Fig. B.2, the breakpoints n^* are chosen where the curves average slope flattens abruptly. n^* increase as the noise is reduced.

A truncated SVD inversion provides a controlled approximation for the spectral function:

$$\tilde{A}^{\text{svd}}(\nu) \simeq \sum_{n=1}^{n_{\text{svd}}} \frac{\tilde{p}_n}{w_n} v_n(\nu), \tag{B.14}$$

The modes higher than n_{svd} are discarded because their coefficients, (which only contribute random noise to the spectral function), blow up exponentially with n. If we know the bound on the signal's convergence rate $|p^{\text{sig}}/w_n|^2 < c\, e^{-\alpha n}$, we can estimate the error in the norm as

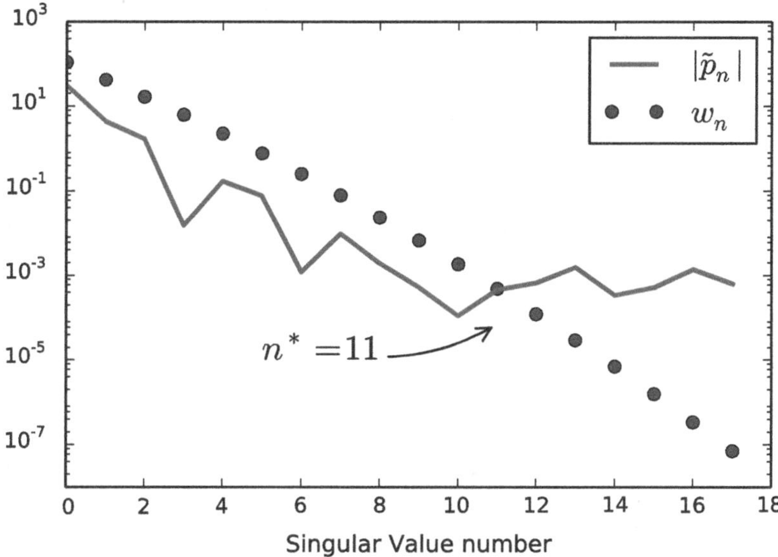

Fig. B.3 Comparison between the projections p_n (*sold likes*) and the singular values of the kernel w_n (*circles*) for the high quality QMC data for the O(2) model. The linear system size is $L = 120$, and coupling constant is $\delta g = 1.17\,\%$. We see that due to the effect of the noise, $\tilde{p}_n$ flattens at the breakpoint at $n^* \approx 11$

$$||\delta \tilde{A}||^2 = \sum_{n=n_{\text{svd}}+1}^{\mathcal{N}} \left| \frac{p_n^{\text{sig}}}{w_n} \right|^2 < \frac{\tilde{p}^2_{n_{\text{svd}}}}{\alpha w^2_{n_{\text{svd}}}}. \tag{B.15}$$

Thus, the smaller the noise level, the larger n_{svd} and therefore the smaller the error in the spectral function, Eq. (B.15).

In Fig. B.3 we show the SVD analysis of the QMC data for the real O(2) model. The projections $\tilde{p}_n$ flatten roughly at $n^* \approx 11$, as they behave in the test model in Fig. B.2.

$\tilde{A}^{\text{svd}}$ can exhibit spurious oscillations due to the missing modes $\{v_n(\nu),\ n > n_{\text{svd}}\}$. This effect, which is part of the error $||\delta\tilde{A}||^2$, is similar to spurious oscillations obtained by a truncated inverse Fourier transform. In cases where it is known that $A(\nu) > 0$, (as for the scalar susceptibility and real conductivity), the SVD truncation can produce unsightly negative regions.

In Fig. B.4 we plot the $\tilde{A}^{\text{svd}}(\nu)$ for increasing values of n_{svd}. We see that indeed the reconstructed solution converges as we increase n and remains stable up to $n \approx 11$, which is where we locate the breakpoint in the SVD analysis. For $n = 12$, the inverted errors dominate the spectral function, which yields a wildly erroneous result.

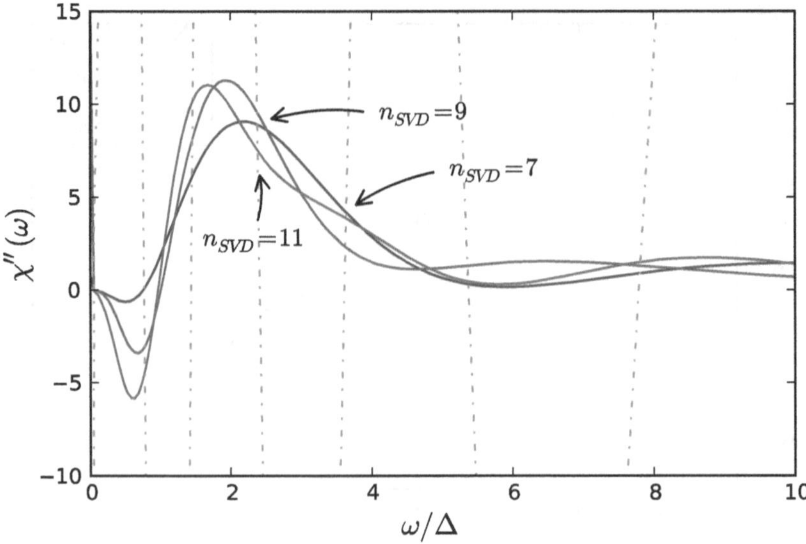

Fig. B.4 Analytic continuation obtained from the first n singular values, for the QMC data of Fig. B.3. We see that spectral functions converge to until $n \leq 11$, in agreement with assigning $n^* = 11$ where $\tilde{p}_n$ starts to flatten in Fig. B.3. For $n_{\mathrm{svd}} = 12$, (*dashed line*) the condition $\tilde{p}_n < w_n$ is violated, and the resulting spectral function wildly differs from the converged function, since it is dominated by random amplified errors

B.3 Maximum Entropy and other Regularizations

The QMC simulation produces noisy variables $\mathcal{G}(i\omega_m)$, whose covariance matrix is defined as

$$\Sigma^{-1} = \big\langle \mathcal{G}(i\omega_m)\, \mathcal{G}(i\omega_n) \big\rangle. \tag{B.16}$$

A condition for the inverted spectral function is that

$$\chi^2 = (\mathcal{G} - KA)^T\, \Sigma\, (\mathcal{G} - KA) \approx \mathcal{N}, \tag{B.17}$$

where $\mathcal{N}$ is the number of data points.

As we have seen before, since K has very small SVD eigenvalues, there is a large family of functions $A(\nu)$ which have the same $\chi^2/\mathcal{N} \approx 1$. The SVD truncation is one way to choose among these functions, but the result may have spurious oscillations, and turn negative in some regions. To improve on this approximation one needs to impose extra conditions on $A(\nu)$, which amounts to extrapolation of Eq. (B.14) to include higher SVD modes. A common approach, which ensures positivity, is to introduce a cost functional $f(A)$, and to variationally minimize

$$Q = \frac{1}{2}\chi^2 + \lambda f(A) \tag{B.18}$$

with respect to A. This minimization lifts the degeneracy in χ^2, and depends critically on the choice of λ. λ can be chosen by the L-curve method [3], which is analogous to the determination of the breakpoint n^* described above.

Two cost functions are commonly used. (1) The 'Maximum Entropy' (MaxEnt) [7],

$$f^{\mathrm{MaxEnt}}(A) = -\sum_i A(\nu_i) \ln A(\nu_i) \tag{B.19}$$

which is based on a Bayesian statistics, and (2) the 'Laplacian',

$$f^{\mathrm{Lap}}(A) = \sum_i \left.\frac{d^2 A(\nu)}{d\nu^2}\right|_{\nu=\nu_i} \tag{B.20}$$

which penalizes unsmooth spectral functions (or long real-time decay). In these functionals, the real frequency ν is discretized as a finite sequence ν_i.

A different strategy is the stochastic regularization [4, 5]. In this method the spectral function is obtained by averaging over a large sample of randomly-chosen solutions consistent with $\chi^2/\mathcal{N} \approx 1$. First a random positive spectral function is generated. Then the goodness of fit is minimized using the steepest decent method while imposing positivity at each step. This procedure is repeated until $\chi^2/\mathcal{N} \approx 1$. Averaging over the random initial conditions leads to the final spectral function.

A complementary approach is to estimate the pole structure of $A(\nu)$, using a Padé approximation. $\tilde{\mathcal{G}}$ is fitted to a rational function

$$\tilde{\mathcal{G}}(i\omega_m) = P_{n_p}(i\omega_m)/Q_{n_p}(i\omega_m), \tag{B.21}$$

where P_{n_p} and Q_{n_p} are polynomials of order n_p. Since $\tilde{\mathcal{G}}$ is an analytic function of $i\omega_m$ we can perform the analytic continuation explicitly by taking $\tilde{A}(\omega) = \mathrm{Im}\,\tilde{\mathcal{G}}(i\omega_m \to \omega + i0^+)$. For the best inversion, one can increase the value of n_p until $\chi^2/\mathcal{N} \approx 1$. Further increase of n_p leads to over fitting and the appearance of spurious poles. n_p needs to be determined, with similar considerations to those determining n_{svd}.

In Fig. B.5 we show a comparison of the different regularization approaches for the same QMC data as used in Fig. B.4. We note that the position of the Higgs peak varies only slightly between different analytic continuation methods, but functions differ somewhat in the higher frequency structure.

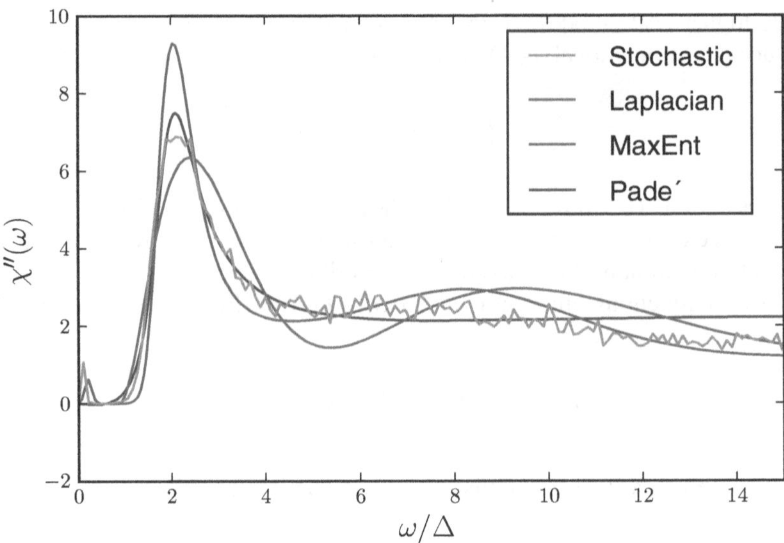

Fig. B.5 Comparison of different regularization methods. Note that the Higgs peak position varies only slightly between the different methods, described in Sect. B.4

As a final note we comment on the form the kernel $K(i\omega, \nu)$ for QMC simulation with discretized imaginary time axis. In this case the imaginary time axis gets a discrete set of values $\tau_i = \Delta\tau \times i$ with $i \in \{0, \ldots, M-1\}$ with $\Delta\tau = \beta/M$. The corresponding Matsubara frequencies are $\omega_m = 2\pi n/\beta$ with $m \in \{0, \ldots, M-1\}$. The kernel is given by a sum over all aliases of the original kernel:

$$\begin{aligned} \tilde{K}(i\omega_m, \nu) &= \frac{1}{\pi} \sum_{k=-\infty}^{\infty} \frac{2\nu}{\left(2\pi(n+Mk)/\beta\right)^2 + \nu^2} \\ &= \frac{\beta}{M\pi} \cdot \frac{\sinh(\beta\nu/M)}{\cosh(\beta\nu/M) - \cos(2\pi m/M)} \,. \end{aligned} \tag{B.22}$$

B.4 Spherical Averaging

A desirable feature of the Eq. (1.3) and its discretized approximation Eq. (2.9) is the Euclidean spacetime symmetry. As a consequence, it is not necessary to single out any one specific direction as the "time" direction. In particular, ignoring weak anisotropies arising from the underlying cubic lattice, correlation functions such as that in Eq. (2.10) are spherically symmetric and only depend on the Euclidean distance from the point $\mathbf{r} = (\tau, x, y)$ to the origin. This is especially correct near the QCP, where the large correlation length ensures that the correlation function at long distances is insensitive to the discrete nature of the lattice.

This observation suggests that one may reduce the statistical noise by performing a spherical average over all possible time directions. In this method, the correlation function at time τ is obtained by averaging over all the points within a thin spherical shell between radius $r = \tau$ and $r = \tau + \delta\tau$. This leads to a large enhancement in statistics—for a $L \times L \times L$ system, $\mathcal{O}(L^3)$ data points are used instead of the $\mathcal{O}(L)$ points typically used in computing the correlation function. In order to implement this method accurately it is necessary to account for the weak anisotropy arising from the underlying cubic lattice. This is done by projecting out the lowest cubic anisotropies prior to the averaging.

The bulk of the data presented in this paper was obtained by averaging over the three principal axes only, and not taking advantage of the full spherical averaging. However, preliminary numerical tests show that spherical averaging does indeed yield high quality results while requiring shorter simulations. This effect may be significant in light of the high sensitivity of numerical analytic continuation to noise. We intend to develop this strategy further in future work.

References

1. M. Jarrell, J.E. Gubernatis, Bayesian inference and the analytic continuation of imaginary-time quantum Monte Carlo data. Phys. Rep. **269**(3), 133–195 (1996). ISSN: 0370–1573, doi:10.1016/0370-1573(95)00074-7. http://www.sciencedirect.com/science/article/pii/0370157395000747. Accessed 28 Mar 2012
2. G.A. Kalyabin, An asymptotic formula for the minimum eigenvalues of Hilbert type matrices. Funct. Anal. Appl. **35**(1), 67–70 (2001). (Bibliography 80)
3. P.C. Hansen, D.P. O'Leary, The use of the L-Curve in the regularization of discrete Ill-posed problems. SIAM J. Sci. Comput. **14**(6), 1487–1503 (1993). doi:10.1137/0914086. http://epubs.siam.org/doi/abs/10.1137/0914086
4. A.S. Mishchenko et al., Diagrammatic quantum Monte Carlo study of the Frohlich polaron. Phys. Rev. B **62**(10), 6317–6336 (2000). doi:10.1103/PhysRevB.62.6317. http://link.aps.org/doi/10.1103/PhysRevB.62.6317
5. A.W. Sandvik, Stochastic method for analytic continuation of quantum Monte Carlo data. Phys. Rev. B **57**(17), 10287–10290 (1998). doi:10.1103/PhysRevB.57.10287. http://link.aps.org/doi/10.1103/PhysRevB.57.7

Appendix C
Complex Conductivity

In this section we will derive the complex conductivity for the disordered and ordered phase in weak coupling.

To one loop order, the paramagnetic response in the disordered phase in given by [1]:

$$\Pi^{\mathrm{P}}_{xx}(p) = \int \frac{d^3q}{(2\pi)^3} \frac{4q_x^2}{q^2+\Delta^2} \cdot \frac{1}{(q+p)^2+\Delta^2} \tag{C.1}$$

where Δ is the renormalized single particle gap in the disordered phase and $p = (\omega_m, 0, 0)$. Introducing the Feynman parameter x and shifting $q \to q - xp$,

$$\Pi^{\mathrm{P}}_{xx}(p) = \int_0^1 dx \int \frac{d^3q}{(2\pi)^3} \frac{4q_x^2}{\left[q^2 + x(1-x)p^2 + \Delta^2\right]^2} . \tag{C.2}$$

Performing the q integration up to a cutoff Λ we obtain:

$$\Pi^{\mathrm{P}}_{xx}(p) = \frac{2}{3\pi^2} \int_0^1 dx \left[\Lambda - \frac{3\pi}{4}\sqrt{p^2x(1-x)+\Delta^2}\right] \tag{C.3}$$

up to corrections that vanish as $\Lambda \to \infty$. To obtain the full response we must subtract the diamagnetic part. Since the superfluid stiffness vanishes in the disordered phase, this is given by $\Pi^{\mathrm{D}}_{xx} = \Pi^{\mathrm{P}}_{xx}(p \to 0)$. This cancels the linearly divergent term, to yield*-4pt

$$\Pi_{xx}(p) = -\frac{1}{2\pi} \int_0^1 dx \left[\sqrt{p^2x(1-x)+\Delta^2} - \Delta\right] \tag{C.4}$$

$$= \frac{\Delta}{4\pi} - i\frac{4\Delta^2+p^2}{16\pi p} \ln\left(\frac{2\Delta - ip}{2\Delta + ip}\right) . \tag{C.5}$$

S. Gazit, *Dynamics Near Quantum Criticality in Two Space Dimensions*, Springer Theses, DOI 10.1007/978-3-319-19354-0

We analytically continue by taking $p \to -i\omega + \epsilon$, resulting in

$$\Pi_{xx}(\omega) = \frac{\Delta}{4\pi} + \frac{4\Delta^2 - \omega^2}{16\pi\omega} \ln\left(\frac{2\Delta - \omega - i\epsilon}{2\Delta + \omega + i\epsilon}\right) . \tag{C.6}$$

The conductivity is then

$$\begin{aligned} \sigma(\omega) &= \frac{1}{i\omega}\Pi_{xx}(\omega) = \mathrm{Re}\sigma(\omega) + i\,\mathrm{Im}\,\sigma(\omega) \\ &= \frac{1}{\omega}\,\mathrm{Im}\,\Pi_{xx}(\omega) - \frac{i}{\omega}\mathrm{Re}\Pi_{xx}(\omega) \end{aligned} \tag{C.7}$$

We note that $\mathrm{Re}\sigma(\omega)$ vanishes for $\omega < 2\Delta$. Above the threshold we obtain:

$$\mathrm{Re}\sigma(\omega) = \frac{\omega^2 - 4\Delta^2}{16\omega^2} \qquad (\omega > 2\Delta) . \tag{C.8}$$

The imaginary part is given by:

$$\mathrm{Im}\,\sigma(\omega) = \frac{1}{16\pi\omega^2}\left[(4\Delta^2 - \omega^2)\ln\left|\frac{2\Delta - \omega}{2\Delta + \omega}\right| + 4\Delta\omega\right] \tag{C.9}$$

In the ordered phase the paramagnetic response is given by [2. 3]:

$$\Pi^{\mathrm{P}}_{xx}(p) = \int \frac{d^3q}{(2\pi)^3}\frac{4q_x^2}{q^2 + m^2} \cdot \frac{1}{(q+p)^2} . \tag{C.10}$$

Here m is the Higgs mass. As before we introduce the Feynman parameter x and shift $q \to q - xp$:

$$\Pi^{\mathrm{P}}_{xx}(p) = \int_0^1 dx \int \frac{d^3q}{(2\pi)^3} \times \frac{4q_x^2}{\left[q^2 + (1-x)(xp^2 + m^2)\right]^2} . \tag{C.11}$$

Performing the q integration we obtain:

$$\Pi_{xx}(p) = \rho_s + \frac{m(3m^2 + 5p^2)}{24\pi p^2} - i\frac{(p^2 + m^2)^2}{16\pi p^3}\ln\left(\frac{m - ip}{m + ip}\right) . \tag{C.12}$$

In the final expression we absorbed the constant term (including the linear divergence) and the diamagnetic contribution into the superfluid stiffness ρ_s.

After analytic continuation the real conductivity is given by:

$$\mathrm{Re}\sigma(\omega) = \pi\rho_s\delta(\omega) + \frac{(\omega^2 - m^2)^2}{16\omega^4}\,\Theta(\omega - m)\,, \tag{C.13}$$

with ρ_s being the superfluid stiffness. The imaginary part of the conductivity is

$$\begin{aligned}\mathrm{Im}\,\sigma(\omega) = \frac{-\rho_s}{\omega} + \frac{(m^2 - \omega^2)^2}{16\pi\omega^4}\ln\left|\frac{m - \omega}{m + \omega}\right| \\ + \frac{m(3m^2 - 5\omega^2)}{24\pi\omega^3}\end{aligned} \tag{C.14}$$

References

1. K. Damle, S. Sachdev, Nonzero-temperature transport near quantum critical points. Phys. Rev. B **56**(14), 8714–8733 (1997). doi:10.1103/PhysRevB.56.8714. http://link.aps.org/doi/10.1103/PhysRevB.56.8714
2. N.H. Lindner, A. Auerbach, Conductivity of hard core bosons: a paradigm of a bad metal. Phys. Rev. B **81**(5), 054512 (2010). doi:10.1103/PhysRevB.81.054512. http://link.aps.org/doi/10.1103/PhysRevB.81.054512
3. D. Podolsky, A. Auerbach, D.P. Arovas, Visibility of the amplitude (Higgs) mode in condensed matter. Phys. Rev. B **84**(17), 174522 (2011). doi:10.1103/PhysRevB.84.174522. http://link.aps.org/doi/10.1103/PhysRevB.84.174522

Zeitfracht Medien GmbH
Ferdinand-Jühlke-Straße 7
99095 Erfurt, Deutschland
produktsicherheit@kolibri360.de